Women
of the Air

Judy Lomax

DODD, MEAD & COMPANY   *New York*

First published in the United States in 1987
Copyright © 1986, 1987 by Judy Lomax
Published by Dodd, Mead & Company, Inc.
71 Fifth Avenue, New York, N.Y. 10003
Distributed in Canada by
McClelland and Stewart Limited, Toronto
Manufactured in the United States of America
Originally published in Great Britain by
John Murray (Publishers) Ltd. London, 1986.

FIRST EDITION

Library of Congress Cataloging-in-Publication Data

Lomax, Judy, 1939–
  Women of the air.

  Bibliography: p.
  Includes index.
  1. Air pilots—Biography. 2. Women air pilots—
Biography. I. Title.
TL539.L55   1987   629.13′092′2 [B]   86-24206
ISBN 0-396-08980-1

1 2 3 4 5 6 7 8 9 10

Contents

Illustrations

AUTHOR'S ACKNOWLEDGEMENTS

The willing support of my family and friends has been invaluable, as has the help given by the many new friends made during research for this book. In particular, I should like to thank John Coleman and Marie Thomas at BBC Southampton; Peter Skinner and Ann Wood in America; Elfrieda Chudoba in Austria; Molly Sedgwick, Lettice Curtis, Joan Hughes, Diana Barnato Walker, Lady Rosemary du Cros, Judith Chisholm, Sheila Scott and Jim Reynolds in England. I am also grateful to the many people, both at home and abroad, who have been willing to answer telephone calls and letters, and who have gone out of their way to offer advice and information. I hope that all these people will enjoy reading this book as much as I have enjoyed writing it and meeting the 'women of the air' either in person or through their own and other people's written and oral accounts.

Introduction

Aviation was created by men, and its history is often told as a chronicle of the achievements of men. They attempted to emulate the flight of birds by jumping off towers or flapping man-made wings long before 1709, when a Brazilian priest, Bartolomeu de Gusmão, demonstrated the first hot-air balloon, a model flown indoors in Portugal. All but a small minority of cranks and enthusiasts still agreed with the Frenchman, Josephe de Lalande, who later in the eighteenth century stated categorically that 'There is no means by which a human being could lift himself into the air and float above the earth.' Within a few years, the Montgolfier brothers in France had proved him wrong by using hot air to inflate a balloon capable of carrying passengers, and another Frenchman, Professor Jacques Charles, had followed their experiments with a balloon filled with hydrogen. The possibilities of floating flight were limited, but it was not until the middle of the nineteenth century that the powered airship was developed from the balloon, and another thirty years before there was any directional control over airships.

At last, in 1903, the Wright brothers in America discovered how to sustain controlled flight in an aircraft powered by an engine. Seven years later, the Frenchman Louis Blériot made an aerial sea crossing in a powered aircraft, flying across the English Channel in a monoplane. Aircraft piloted by men were used in the First World War, and afterwards men were the first to pioneer new routes, flying further and further in aircraft which at first were still flimsy wooden constructions with open cockpits.

Flying in the early days was hazardous, and the pioneers of aviation needed courage and determination, qualities shown by the women who, although in a minority, followed the men into the air, where their presence was often resented. Their successes made them instant heroines, their failures were used to prove that women were physically and psychologically unfit to fly, and their survival as proof of the safety of aviation.

To tell the story of the women who, from the early days of ballooning, have defied convention and fought against prejudice by taking to the air is not to denigrate the men. The great male pioneers of aviation deserve to be honoured and remembered. So do the women who followed, competed with, and sometimes led them, proving that women could fly as well as men, could endure the same hardships and earn the same recognition.

1

Early Aeronautics
and Edwardian Entertainment

The first passengers in a balloon were neither men nor women, but a cockerel, a sheep and a duck. On 19 September 1783 this unfortunate menagerie was sent two miles high by the Montgolfier brothers. Watched by 30,000 people, including King Louis XVI and Queen Marie Antoinette, they survived the ordeal, although the sheep apparently either trod or sat on the cockerel. They made a somewhat hasty descent after eight minutes when the balloon crashed into a tree and the rope holding their basket broke.

As it was considered so dangerous and unnatural to leave the ground, the king at first insisted that manned balloon flight tests should be conducted only with criminals. He was eventually persuaded to allow Pilâtre de Rozier, one of the Montgolfiers' assistants, to make first a tethered ascent and then, with the Marquis d'Arlandes, the world's first free flight.

The first recorded female tethered ascent was also made in France in May 1784, by four women, three of them titled. A month later Madame Elisabeth Thible became the first woman to fly in an untethered balloon as a passenger, watched by the king of Sweden. Madame Thible, who flew a mile high, elegantly attired in a lace-trimmed dress and a feathered hat, did not repeat the experience, although she apparently enjoyed it so much that she burst into song.

Less than a year after the first manned flight in France, a balloon voyage, of 24 miles in two-and-a-half hours, was made in Britain in September 1784 by Vincent Lunardi, secretary to the Neapolitan ambassador in London, with a pigeon, a cat and a dog: the pigeon flew away and the cat escaped when he made a brief touch-down on the way. Lunardi's bonnet and garters instantly became fashionable with his many female admirers, one of whom he invited to join him and a certain Mr Biggin on a flight. Mrs Letitia Ann Sage was both the first Englishwoman to go up in a balloon, and the heaviest person so far in the air: because of her great weight, the pilot had to relinquish his place. However, the two novices went up, and came down again, safely without him. Although she does not appear to have made a habit of ballooning, Mrs Sage remarked afterwards that she was 'infinitely better pleased than I ever was at any former event of my life'.

The first female ascent from England had been made the previous month by a fourteen-year-old French girl, a Mademoiselle Simonet, with Jean-Pierre Blanchard, France's leading showman of the air. Blanchard toured in Europe and America, drawing crowds by dropping cats and dogs on parachutes: on one occasion a cat, a dog and a squirrel shared a single parachute.

Madeleine Sophie Blanchard, who became the first woman aeronaut in her own right, made her first ascent in 1805. The Blanchards were an unprepossessing couple. He, according to an uncharitable contemporary account, 'was a petulant little fellow, not many inches over five feet, and physically well suited to vapourish regions'. His wife was 'small, ugly and nervous', terrified of riding in a horse-drawn carriage, and hated noise so much that she liked to spend the night in a balloon. In spite of her phobias on the ground, she continued to make a living from ballooning after her husband's death and gained such a reputation that Napoleon made her Official Aeronaut of the Empire.

Madame Blanchard was killed in July 1817 during an aerial firework display billed as the main attraction at an evening fête at the Tivoli Gardens in Paris. These displays were her speciality, and after her ascent, accompanied by music, it took a few minutes for the applauding crowd to realise that they were watching not aerial fireworks but a fire in the balloon itself: the hydrogen had caught alight and the balloon was rapidly losing height. Madame Blanchard no doubt had a plentiful supply of the miniature parachutes to which her fireworks were attached, but neither she nor her husband had ever investigated the use of the parachute as a means of escape. Her attempt to regain height by throwing out ballast failed: when the balloon hit a rooftop, she was thrown out and died of a broken neck.

The first parachute to take the weight of a human body was

designed by a Frenchman, André Jacques Garnerin. He had
dreamt of such a way of escaping from a fortress when he was
imprisoned during the French Revolution. Garnerin demonstrated his parachute first in France in 1797, and then in England
in 1802 when, because of its violent oscillation, he gained the
doubtful distinction of being the first person on record to be
airsick. He later amended the design, and in 1815 his wife became
the first woman to parachute – although once was enough. Their
niece, Eliza Garnerin, became the first professional parachutist,
giving thirty-nine parachuting displays between 1816 and 1836.
Later in the century her compatriot, Madame Poitevin, specialised in parachute descents – that is, when she and her husband
were not entertaining the crowds by making mounted equestrian
balloon ascents. This particularly cruel French sport was started
in 1798 by one Testu Brissy, continued by Margat who ascended
on the back of a stag, and also by both Monsieur and Madame
Poitevin. When the Poitevins visited England in 1852, the public's hatred of cruelty to animals succeeded in banning her from
going up as Europa on the Bull; but no objection was raised to her
risking her own neck by parachute drops.

Not until nearly the end of the nineteenth century, however,
did parachuting become more than an occasional addition to
ballooning as an entertainment in England. In the 1880s Park van
Tassel, an American of Dutch descent, designed a limp canvas-
and-rope bag which would be forced open and fill with air as soon
as it took the weight of a falling human body. Instead of a basket
for the parachutist, there was a trapeze from which he – or she –
could hang, supported only by a sling. Van Tassel's parachute
was copied by an American wire-walking acrobat, Thomas Scott
Baldwin, who brought it to Alexandra Palace, already a
flourishing ballooning centre. Baldwin made twelve descents in
1888, spread over two months, during which time heated con-
troversy in the press and even in the House of Lords ensured the
widest publicity. The eight-year-old Balloon Society gave Baldwin a gold medal.

Ballooning displays and joy-rides featured prominently in
British Victorian outdoor entertainment, and for a few years after
the Aero Club was founded in 1901 balloon meets at Ranelagh
and Hurlingham were almost as fashionable as Ascot, but
although colourful and picturesque, ballooning as a spectator

sport lacked the element of suspense so dear to the showman's pocket. This element was added by the men and women who were prepared to risk their lives and thrill the public by jumping with parachutes from a great height. The sky was considered a place of great mystery and hazard, and jumping into it from a balloon the greatest hazard of all. Fête organisers and showmen made the most of such spectacular descents by parachute from both hot-air and gas balloons; they quickly realised that 'an attractive female face and a trim figure drew more spectators to an aerial event than even the most dashing male parachutist'.

Parachuting from a balloon was still dangerous enough to provide the occasional drama, however, and in 1895 an August Bank Holiday fête ended in tragedy. The headline in the *Peterborough Advertiser* announced 'Awful Fatality to a Lady Parachutist at Peterborough, a Bank Holiday Sensation'. Ten thousand people were watching 'the comely and brave Mademoiselle Adelaide Bassett and by her side Captain Orton, both balloonists of experience and some renown' who were to ascend by balloon and then make simultaneous parachute jumps and 'gracefully descend to earth'; 'Alas the throng of spectators witnessed a thrilling tragedy which was to send them homewards horrified and which was calculated to touch the stoutest heart. The balloon rose and struck a tree. Wires detached the Mademoisell's [*sic*] parachute. She jumped and fell 200 ft to earth.' Her death nevertheless did not deter another couple, Captain Spencer and Miss Alma Beaumont, from making a double parachute jump a couple of days later.

Captain Spencer was one of a ballooning family whose activities were based at Alexandra Palace, where the sixteen-year-old Dolly Shepherd went one spring day in 1903 to hear the music of John Philip Sousa. Dolly Shepherd was a determined young lady and, as she could not afford a ticket, she talked her way into a temporary job as a waitress. Among the first people she had to serve were John Henderson, Auguste Gaudron and Samuel Cody. John Henderson was the director of entertainment at Alexandra Palace, or 'Ally Pally' as it was popularly known, where he had the task of co-ordinating the endless round of pleasure provided at the racecourse, fairground, boating lake, theatres, arenas and enclosures where carnivals, parades, demonstrations and firework displays were held. Auguste Gaudron, 'a dapper little man' with a soft French accent, clear piercing eyes and a neat waxed moustache, was a parachutist and balloonist whose aerial skills were much in demand; while the cowboy stunts of Samuel Franklin Cody, a long-haired Wild West show-man as flamboyant as his namesake W.F. 'Buffalo Bill' Cody, financed his more serious aim of creating the first kite capable of carrying a man through the air.

Samuel Cody's entertainment 'The Klondike Nugget', a stage spectacular using the skills that he had learnt as a Texas cowboy, was currently drawing crowds to Alexandra Palace at the Bijou Theatre. His wife, Lela, was the centrepiece of one of the acts, in which he shot a plaster egg from her head after being blindfolded by their son. In spite of his skill and Lela's confidence in his ability, one evening his aim was less than perfect, and the bullet grazed her scalp. Dolly Shepherd promptly offered to act as stand-in. Cody repaid her with the greatest compliment he could imagine: an invitation to visit the aeronauts' workshop which filled the banqueting hall, and permission to sit in the wicker seat which hung beneath one of his enormous box kites, a privilege normally granted only to his wife, whom he had once left suspended in mid-air for hours while he became engrossed in some minor adjustment.

Dolly was introduced to Henry Spencer, one of the three aeronautical brothers whose father and grandfather had also been leaders in the ballooning fraternity, and was then handed over to Gaudron to continue her tour among the balloons which were heaped or spread over much of the floor. She was enthralled by the wicker baskets, netting, rope, cord, and parachutes that hung limply from the ceiling, and by the combined smells of glue, canvas, hemp and dust and the background clatter of sewing machines. Finally Gaudron capped her stream of questions with one of his own: would she like to make a parachute descent – to which her answer was an immediate and enthusiastic 'yes'.

Gaudron, already a professional balloon-maker when he came to England from France at the age of twenty-two, had been performing at Alexandra Palace since 1898. With the Short brothers and the Spencer brothers, he was among the top balloon manufacturers in England, and his friendly rivalry with the Spencers was sealed by his marriage to their sister. With Henry Spencer he was working at Alexandra Palace on a cylindrical 'dirigible', over which the shape allowed some directional control, for Dr Barton, with whom he had dropped the first airmail letters over Britain in 1902 from a special Coronation balloon. He was best known, however, as the leader of a team of stunt parachutists, which he invited Dolly to join a year later after one of the girls, Maud Brooks, had been severely injured in a parachuting accident in Ireland.

Far from being worried by such danger, Dolly was delighted to accept the invitation, and Gaudron, after squeezing her hands and asking her to hang from a trapeze for a few seconds, declared her physically strong enough: that she was not of a nervous disposition he already knew from her appearance in 'The Klondike Nugget'. During her training, which lasted half an hour, she was shown how to hold on to the trapeze bar and step into the small canvas sling which would pull up between her legs and relieve her arms of her body weight. A safety belt was considered an optional extra, used only in exceptionally windy weather and as likely to cause as to prevent injury. The most important lesson was how to land: rather than falling sideways, as is done now, it was then considered that, to avoid injury to the spine, it was essential to roll over on to the back and throw the legs up in the air immediately the feet touched the ground.

The long skirts normally worn by young ladies at that time were clearly unsuitable for such antics, so Dolly's uniform was a waisted navy blue knickerbocker suit with gold trimming, long front-laced boots, and a high peaked cap. On the lapel of her suit was a badge showing a parachute, and on the cap a balloon insignia. Dolly considered her appearance in this outfit fetching, but her enthusiasm was not shared by her aunt Mariam, with whom she was living in London and by whom she was employed at the Ostrich Feather Emporium in Holborn. Although Dolly's parents raised no objections, her aunt disapproved strongly of what she termed 'mountebank stunts', an attitude which took Dolly by surprise: to her, parachuting 'just seemed like an exciting and challenging adventure . . . Rather daring and unconventional, yes – but surely not that bad?'

It was not long before Dolly became an accepted and popular member of the team, working her way rapidly from a beginner's jump from the edge of the balloon's passenger basket, to an ascent swaying beneath the basket from which she then detached herself, and then to a solo descent with no basket and no one to give advice. When she had decided where to land, she pulled the ripping cord to detach herself from the balloon, which then deflated rapidly and landed unattended. Her costume showed off her trim figure and curly hair, and she enjoyed the admiring glances her appearance attracted, especially from the gentlemen, while she mingled with the crowd before the ascents. Elegantly dressed in a variety of long-skirted outfits with matching gloves and hat, she would answer awed questions with aplomb, chat equally easily with the humblest and the most elevated members of society, and make friends everywhere.

The ascent, a peaceful dream-like experience, with no sensation of movement, was followed by the exhilaration of the descent, plummeting at first like a stone until, with a rapid flapping of silk sucking at the rush of air, at last there was a big *whooosh* as the parachute opened, the sling tightened, and the trapeze bar tugged at her arms. The most difficult part of any jump, however, was the landing. As the ground approached, there was always the danger that a sudden gust of wind might blow the parachutist off course, and trees and bushes which looked insignificant from several thousand feet up would assume alarming proportions. Dolly landed in a tree only once, and once on a roof, but another member of the team, Viola Kavanagh, was killed after landing on a factory roof when the wind lifted her and dashed her to the ground.

Usually all the team members made a thorough check of their equipment before each ascent, although even this could not prevent something occasionally going wrong. When the cotter pin attached to the ripping valve jammed on one solo flight, Dolly was carried swaying from the trapeze bar for several hours. To keep her strength and courage up, she sang everything she could think of and, by the time the balloon at last touched down for long enough to free her, long after dark, she lay exhausted on the ground as it floated on. It was found three days later, with the parachute still attached, in the North Sea.

On the only occasion when Dolly neglected her usual careful inspection of her parachute, she had arrived late at a fête where she and 'Captain' Gaudron were to make a double descent from a 'right-away' balloon – one which, after the jump, continued on its way to land later with its fare-paying passengers. After a cursory glance at her parachute, she stepped into the sling, grasped the trapeze bar, and was whisked into the air beneath the balloon basket. As she swung over the crowd, she heard cries of horror instead of the usual cheering and was aware of a new and alarming sensation; somehow the parachute had become detached from the balloon, and she was making a completely free fall into the crowd. The people on whom she fell cushioned her landing and, although one woman fainted, they and Dolly escaped with no more than severe bruising.

The mystery of how the cords, part of which still dangled beneath the balloon basket, had come to break was never solved. Dolly had no chance to discuss her suspicion of sabotage with Gaudron, for he too came to grief. It was a gusty day, and the wind dashed him first into a brick wall as he landed and then, still attached to his parachute by his safety belt, which had no means of quick release, he was dragged over one flint wall and into another. He suffered temporary loss of sight in one eye and injuries to his face and neck. The headlines next day were dramatic: 'PARACHUTISTS ESCAPE – DOLLY SHEPHERD NARROWLY MISSES DEATH – CAPTAIN GAUDRON SERIOUSLY INJURED'.

Her most serious accident left her semi-conscious for several days, and for some weeks afterwards it seemed unlikely that she would ever walk, let alone parachute, again. Although forbidden to discuss her exploits with her aunt, with whom she was still living, Dolly was never naturally inclined to reticence, and the girls who worked an eleven-hour day under her management at the Ostrich Feather Emporium were as eager to hear the details of her jumps as she was to tell them. Even though she was a strict disciplinarian, and once even carried out a threat to push one of the girls into a vat of dye, Dolly's stories brought glamorous relief to the tedium of their work.

One of her employees, Louie May, longed to share Dolly's parachuting experiences, although nothing would have induced the rest to do anything so dangerous. At Dolly's suggestion, Gaudron interviewed Louie, found her suitable, gave her the standard half-hour's training and put her on his reserve list. Her opportunity came when Dolly was double-booked; Louie, in the strictest secrecy as both her fiancé and her grandmother with whom she lived would have forbidden it, was to make her first jump from the basket of a 'right-away' balloon. But the weather decided otherwise: it was far too windy for a beginner's jump.

To make up for both Louie May's and the crowd's disappointment, it was decided that she and Dolly would make a double descent the following day from the basket of Gaudron's *Mammoth*, the largest balloon so far built in Britain. Gaudron had recently set a hazardous 1150-mile record in *Mammoth*, which had carried him with the wind from Crystal Palace through a snowstorm to the icy bank of a lake in Russia. Once again Louie May was to be disappointed. *Mammoth* was already inflated, 'glistening in the sunlight and standing supreme, like a resplendant goddess with her subjects clustered about her', when because of a damaged valve the gas started to escape and the balloon to sag. There was only one way for the show to go on: the two girls would make a double descent from the much smaller balloon which Dolly usually used for a solo drop. A ring and cotter pin were improvised to attach Louie's parachute to the balloon.

The ascent went smoothly. Louie was as entranced as Dolly by the exhilaration, the stillness and the silence, as the landscape beneath them became smaller and smaller. At 3000 ft, they prepared to pull the cords which would release them and send them floating back to earth. But Louie's pin was jammed, and nothing that she or Dolly could do would free it. Suddenly they were in fog, and then floating under the balloon above a thick layer of cloud. The aneroid read 11,000 ft as Dolly realised that her companion was so frightened, tense and cold that she would not be able to hold on to the trapeze bar much longer.

Although double descents were common, the two parachutists always jumped separately, and Dolly's parachute was designed to take no more than her own weight. The only way down was, however, together. In this, the first mid-air rescue ever attempted, Louie was transferred from one parachute to the other, two miles above the earth with nothing but her and Dolly's determination to succeed between them and death. At last, Louie's legs and arms were around Dolly's waist and neck, her full weight hanging on Dolly as she pulled the ripping cord. It was hardly the initiation to parachuting that Louie had hoped for as they hurtled together through the air, waiting for the parachute to fill above them.

Inevitably their descent was too fast and their landing heavy, with Louie on top of Dolly as they hit the ground. Their initial reaction, once Louie had been persuaded that she had not lost all her teeth, was to laugh hysterically. Then, although she could feel nothing, Dolly realised that she was unable to get to her feet. She spent the next few weeks at a nearby farmhouse while argument raged in the local paper, the *Staffordshire Sentinel*, about whether parachuting stunts should be allowed to continue. When she discovered that she was not expected to walk again, Dolly was in despair until a local country doctor, well in advance of the specialists, used a battery and a volunteer to form an electrical circuit and gave her electric shock treatment. This brought feeling back into her legs and back and, with the unstinting care of the Holland family, who had taken her into their farmhouse, she was soon fit again and ready to take to the air.

Eight weeks after the accident, Dolly fulfilled an engagement at Ashby de la Zouch. Although she was apprehensive, as soon as her feet left the ground and she was hoisted above the wildly cheering crowd, her fears evaporated: she was still the 'Parachute Queen'. Louie May, not surprisingly, never made another parachute jump. During Dolly's convalescence at the farm, stand-ins had taken her place at the shows for which she had taken advance bookings, among them her mother, who was billed at a show in the London area as Madame Papillon. It was a secret she guarded so closely that Dolly did not learn of it until 1945, shortly before her mother's death.

Dolly Shepherd made her last jump on a perfect spring day at Alexandra Palace in 1912. As usual, she mingled first with the crowd, and waved a Union Jack as, alone, she soared and swayed over the watching people. In the silence, she heard a warning voice: 'Don't come up again, or you'll be killed.' She was twenty-five and wanted to live. Several of her fellow parachutists in Gaudron's team were now dead: 'Captains' Smith and Fleet as well as Viola Kavanagh. For eight years, she had enjoyed life to the full, was treated as a celebrity, but was always as astounded as after her first jump to find herself the object of adulation. She had, after all, only set out to enjoy herself: 'The best hotels, fine meals, excellent entertainment, the attendance of numerous gentlemen, and the applause of the crowds – all this and parachuting too! Can you wonder that I enjoyed those days?'

Within two years of the 'Parachute Queen's' retirement, the days of carefree aerial entertainment were over. During the First World War Dolly Shepherd was a driver-mechanic in France. She married a Captain Sedgwick who was one of her passengers, and, in the Second World War, became a Shelter Staff Officer, her presence in a shelter always somehow removing its occupants' fears. A few years before her death in 1983, at the age of ninety-six, she flew with the Red Devils, fascinated by their modern parachuting techniques and particularly pleased to meet Jacky Smith, the sole woman member.

2

Pioneer Passenger: Gertrude Bacon and the 'Very Beginning'

In April 1903 a white-haired American called Octave Chanute added his signature and a drawing of a blindfolded pig to the visitors' book kept by Gertrude Bacon. Little did Gertrude realise, as she persuaded her father's ballooning friends to add to her collection of pig sketches and listened to Chanute's description of the tests which the brothers Wilbur and Orville Wright had recently carried out with his glider on an American beach, that she had been hearing about the birth of winged flight.

The Reverend John Bacon, self-styled 'scientific aeronaut', liked to think that he was distantly related to Roger Bacon, a Franciscan monk who had claimed in the thirteenth century: 'It is possible to make engines for flying, a man sitting in the midst whereof, by turning only about an instrument, which moves artificial wings to beat the air, much after the fashion of a bird's flight.' He had also suggested that a hollow metal flying vessel filled with 'aetherial air or liquid fire' would float on the surface of the air like a vessel on water. John Bacon's own interest in ballooning dated from his first ascent in 1888: the following year he gave up his part-time ecclesiastical duties because of what he considered the intolerant attitude shown by the church towards science. Ballooning extended the range of his studies in audio-dynamics, meteorology, wireless telegraphy and military reconnaissance, and he was soon financing his aeronautical expeditions and research work through articles and lectures with such titles as 'The Balloon as an Instrument of Scientific Research'.

Gertrude Bacon, born in 1874, was taught by her father to share his multifarious interests: printing, handbell ringing, beekeeping, amateur dramatics, home-made fireworks, flower, cat and donkey shows, cycling, photography, astronomy, and finally ballooning. She had to be content with acting as hostess to his numerous aeronautical associates for ten years before she herself was at last invited to join John Baker and his friend Percival Spencer on a balloon trip. Long before that, however, she had achieved fame, even notoriety, in the quiet community of Cold Ash in Berkshire by riding a bicycle, considered a most unsuitable activity for a girl in spite of the elaborate elastic arrangements which kept her ample skirts demurely over her ankles.

A man with advanced ideas about many things, John Bacon had taken sole charge of his children's education, teaching them French and Latin, a little Greek, mathematics, natural sciences, elementary astronomy, chemistry, botany and physics. Gertrude and her brother learnt to spell on a little printing press installed to print handbills for local shows, and their father built a small observatory in the corner of a field to further the family interest in astronomy. In 1890 Gertrude became, at sixteen, the youngest 'Original Member' of the British Astronomical Association and in her twenties she accompanied her father on three memorable sea trips to observe and photograph eclipses in Norway, India and America.

When, at the age of twenty-four, she was at last summoned by telegram for her first balloon ascent, her excitement was mixed with anxiety about her poor head for heights; she was relieved to discover that looking down from a balloon, and later from an aeroplane, was quite unlike standing on the top of a high building or on a cliff-edge. As she was carried from Crystal Palace across London to Hertfordshire she enjoyed the sensation that it was the earth, rather than the balloon, that was moving. But the landing was less enjoyable: there had been a drought, and the grapnel hooks which should have dug into the ground to anchor the balloon were made useless, so balloon, basket and occupants were bounced wildly across the ground by a stiff breeze towards a railway line and brought to a halt, a few feet from the track, only when the balloon became entangled in overhead cables.

Her second ascent was equally eventful. They started from Woolwich, drifted over fog-bound London, and ended abruptly in a field near Hertford when the balloon's long trail rope caught in an oak tree. It was disentangled by a railway porter, and as the weather had cleared, the balloon party – Gertrude, her father, and Percival Spencer – decided to take off again. But the balloon, damp after the fog, refused to leave the ground in the still air, and they managed to rise only after cutting free the heavy trail rope – which the porter was instructed to 'rail back to London' – and emptying all their sand ballast. Soon they were well above the cloud, and still rising until, with no oxygen, they were three miles high – when they unexpectedly heard a cow mooing, a phenomenon which John Bacon noted for later scientific consideration. Eventually they began to descend more rapidly than they would have wished, and resorted to throwing out everything else they could, including their scientific instruments.

It was with Stanley Spencer, Percival's brother, that the Bacons made their most memorable flight, a night trip in November 1899 from Newbury with a paraphernalia of food, scientific and photographic equipment and blankets. The idea that they might need to escape from the balloon did not enter their heads, and although the Spencer family balloon business also manufactured parachutes they took neither these nor lifejackets. It was, after all, unlikely that the balloon would, after taking off from the Newbury gasworks, come anywhere near the sea.

There was a serious scientific reason for the wet and cloudy nocturnal departure: to see Temple's Comet and its accompanying meteor shower, the Leonids. John Bacon was to report exclusively on the expected spectacular display for *The Times* whose editor, Moberly Bell, he had persuaded to supply the balloon. Gertrude, with her unwieldy plate camera, was to provide photographic evidence. When they left Newbury at 4.30 a.m., Stanley Spencer expected that, with the light wind, they would be taken west for no more than two or three hours and would then land safely well inland of the Bristol Channel 60 miles away. At 1500 ft they were still in dense cloud, but after throwing out four 70-lb bags of sand they reached clear sky above the cloud at 3000 ft.

There was no sign of meteors, Jupiter having unfortunately diverted Temple's Comet, but the peace and beauty of the night above the cloud made up for the disappointment. Gertrude's description was lyrical:

No one but we three, in all the world, saw that wonderful scene that morning . . . In the blue-black velvety sky twinkled, with metallic lustre, the stars, and the full moon shone clear and cold; but her face was of tawny copper hue and round her was a glorious double halo of brightest rainbow tints; while just below us lay that silent ocean of filmy, tossing billows all silvered in the moonlight, stretching wave after snowy wave, with dark blue shadows in their depths, to the limitless horizon – gleaming, ghostly-pure – of heaven itself.

The sounds of the early morning drifted up to them through the cloud: dogs barking, trains, the steam hooter of Westbury Iron Works, a church clock striking six, the dawn chorus of the farms below them. Then, to their increasing alarm, instead of descending slowly the balloon continued to rise for another five hours as the wind took them nearer and nearer the coast. John Bacon wrote his press report and Gertrude took photographs while Stanley Spencer worried, especially when they heard below them first a ship's siren and then waves on a shingly beach.

As much to pass the time as with any serious hope of success, Gertrude wrote forty-eight distress messages on press telegraph forms: 'URGENT – Large balloon from Newbury overhead, above clouds. Cannot descend. Telegraph to sea-coast to be ready to rescue. Bacon and Spencer, 11.15, Thursday.' Folded into cocked hats, these were thrown from the basket: years later, she met a Welsh farmer who had found one but had had no idea what to do about it. It was after midday when they at last began to descend, eventually, after ten hours in the air, making a rough landing only half a mile from the sea on a Welsh mountainside where they were dragged through barbed wire and across the top of an oak tree. Gertrude's arm was broken in the landing, and she fainted for the only time in her life.

Although their other balloon flights were less eventful, Gertrude continued to accompany her father in his scientific experiments on the nature and behaviour of sound both in the air and on, and even under, the ground: they went down the Great Dolcoath Mine in Cornwall, visited the construction work of the tube line which was to link Waterloo and Baker Street under the Thames, spent a night at the top of Tower Bridge, climbed across the roof of the St Pancras Hotel, and spent the night of Edward VII's coronation at the top of St Paul's Cathedral.

The Spencer brothers' aeronautical interests included airships, and through them Gertrude was drawn into the excitement of early aviation. By 1904 Stanley Spencer had attempted, almost but not quite successfully, to fly from Crystal Palace round the dome of St Paul's. In France Santos-Dumont, a wealthy young Brazilian, had by then already flown round the Eiffel Tower with a small airship, although Gertrude's description of his achievement was unflattering: 'the wealthy young Brazilian, almost literally, tied his motor-bicycle on to his balloon.' In August 1904, Stanley Spencer invited Gertrude to become 'the First Woman in the World to Make a Right-away Voyage in an Airship'. Both Percival and Stanley Spencer were to entertain the crowds, one with a balloon and the other with his third airship, but it was a windy day, and Percival's big balloon was able to take off only when the wind dropped in the late afternoon.

At last, anxious not to disappoint the crowd, Stanley decided that although it was still windy he and Gertrude would make their planned flight. The airship had a 5-hp Simms engine which worked an 11-ft propeller, and was steered by a big rudder-sail at the stern, with a ballast rope to direct the head up or down. Pilot and passenger sat in a shallow car in a light bamboo framework 50 ft long, suspended below an 84-ft-long bag inflated with household gas. As the airship was taking off, a gust of wind caught it, carrying it straight for the roof of a marquee. Stanley Spencer reacted fast enough to throw out a large bag of ballast, and although it had for a moment seemed as if both the marquee and the crowd were in danger, the only damage was to a row of flags on the roof of the tent. Gertrude's first powered flight combined 'all the rare beauty of balloon travel, the matchless panorama, the space, the freedom, with the charm of motion, the thrilling sense of life'. The throbbing of the engine and the smell of burnt petrol only added to her enjoyment.

Ballooning, however, had not entirely lost its attraction, and in 1907 she was the guest of the French Aero Club on a 100-mile flight. French ballooning, she discovered, had achieved considerable sophistication, with the unaccustomed luxury of a padded seat in the basket and different coloured ropes for valve and ripping seam. What struck her most was the method of throwing out ballast: whereas the Spencers would pick up a bag by its corners and tip out the contents, the French pilot, M. Levée, weighed the sand precisely and used a special spoon to ladle it out.

The French were convinced that France was also the birthplace of powered flight. Santos-Dumont had flown an odd contraption of several box-kites strung together called *The Bird of Prey* – at least, it had hopped from the ground and flown backwards for something under 100 yards. Louis Blériot, a thirty-five-year-old engineer who in 1901 had managed to fly a model with flapping wings, seemed likely to establish a record for the number of monoplanes – still considered impractical by almost everyone else – he had crashed, and had achieved a short flight with the eighth. The motor-driven biplane had evolved from the box-kite: Léon Delagrange and Henry Farman, born English but natural-ised French, had one each. Rumours that two American bicycle makers had flown in America were thought to be without found-ation.

It was not until 1908, when Wilbur Wright flew in France for over an hour, that the French stopped thinking of the Wright brothers as mere *bluffeurs* and conceded that, although Farman had by then flown a circular mile and Delagrange had stayed up for nine minutes, the Americans rather than the French were leading aviation. With Blériot's cross-Channel flight in July 1909, French honour was restored. In England Samuel Cody succeeded in building and flying first a glider and then, in October 1908, a biplane powered by two 50-hp Antoinette engines. As Cody was an American, the honour of being the first British pilot went instead to Moore-Brabazon, who learned to fly in France and proceeded to win the *Daily Mail* £1000 prize for flying the first circular mile in an all-British machine. In June 1909 the new magazine the *Aero* listed 39 British aeroplanes, many of which never flew, although among the men mentioned as experimen-ters in aviation were several who later became well known: C.S. Rolls, Howard Wright, Alexander Ogilvie, G.B. Cockburn, Handley Page and the Short brothers.

Gertrude followed these developments avidly. She went to the first Aero Exhibition at Olympia in 1909, kept scrapbooks of press cuttings about aviation, attended meetings of the Aeronautical Society, and travelled to the first international aviation meeting, *la Grande Semaine d'Aviation de la Champagne*, held in Reims in August 1909. Many of the leaders of early aviation were there: Blériot, Latham, Farman and his pupil Roger Sommer, Ferber, Delagrange, Paulhan, Fournier and the Comte de Lambert among the French majority; Glenn Curtiss from America; and Cockburn from England. Throughout the week, Gertrude watched the aircraft, some lumbering – like Captain Ferber's clumsy box-kite and Louis Breguet's 'Flying Coffee Pot' (which failed to leave the ground) – and some light and graceful – like Latham's Antoinette monoplane.

Among the highlights of the week were two airships and three aeroplanes in the sky at once, and the race for the Blue Riband of the Air, won at a then astonishing speed of 47 mph by Glenn Curtiss. Farman flew round and round the course to create a world distance record of 112 miles and reached 'the awe-inspiring and perilous height' of 500 ft: there was anxious speculation about what would happen to an engine at 1000 ft. Light relief was provided by the apparent pursuit of the press by the French pilot Lefèbvre – one press photographer fell flat on his face to avoid him – and by the confusion caused by the aircraft among the cavalry.

Determined not to go home without flying, Gertrude persuaded Roger Sommer, who had recently broken Wilbur Wright's endurance record by flying for nearly two-and-a-half hours, to take her up on the last day in a Farman biplane. While she was waiting for her flight, Blériot wrote off yet another monoplane, watched by 25,000 people, and Farman took two passengers crouched like frogs on his biplane's wings for a joy-ride. The two planes of Sommer's aircraft, each 35 ft long, were mounted on a four-wheeled chassis to which were strapped long wooden sledge runners. Although there was no provision for passengers, from her hot and uncomfortable position wedged between Sommer and the radiator Gertrude found the take-off and landing smoother than she had expected. In spite of a near-miss with Farman in another biplane, she felt again, 'rapturously happy, absolutely safe and secure'. That evening, she was toasted in the bar as the first Englishwoman to fly in an aeroplane.

Having experienced 'the glory and the glamour of the very beginning', Gertrude was borne along on 'the great wave of aerial enthusiasm and adventure that was sweeping the world', an avid bystander and occasional participant although it apparently never occurred to her to learn to fly herself. Instead, she lectured and wrote on every aspect of aviation. Not long after the Reims air meeting, she interviewed Cody, who had at last achieved recognition for a 40-mile cross-country flight, and was watching from the press stand at an aviation meeting in Doncaster when his heavy biplane, known as the 'Flying Cathedral' and originally designed to carry an artilleryman and gun, was badly damaged after it turned over on the ground. He nevertheless provided much-needed entertainment later in the week, on a day when flying had been rained off, by signing his British naturalisation papers on the race course, using the town clerk's back as a desk. Along with his American citizenship he discarded his long hair and huge sombrero for a more soberly British appearance. This helped to dispel the impression that he was 'something of a crank' – a reputation enhanced by an earlier Channel crossing in a 'kite-boat': he was in the boat, which was pulled by a kite.

Had she been wealthier, Gertrude might well have followed the example of another British pioneering woman passenger, E. Trehawke Davies, whose private income allowed her to indulge her love of flying by buying a series of Blériot monoplanes in which she somehow survived numerous crashes. One of her many close brushes with death came after an abortive attempt to fly to Berlin from Paris: 'we' – she and her pilot, a Mr Astley – 'lost our way . . . and eventually landed at Bonn after eight hours flying in the teeth of a powerful headwind.' On the way back to England, the pilot put one of his feet through the Blériot's floorboards, which had rotted because of its 'irritating habit' of spitting lubricant oil all over the floor. As they eventually landed, burying one wing 18 in. into a beetroot field, they were showered with petrol and oil cans, luggage, a biscuit tin and a Kodak camera, but neither was any the worse for the adventure and their would-be rescuers found them sitting happily munching biscuits. Astley nevertheless was killed not long afterwards in a crash in Ireland, where his intrepid passenger would have been with him had she not been busy negotiating the purchase of a new aircraft. She subsequently made several Channel crossings with Gustav Hamel, on one of which they set a record of 12½ minutes.

Gertrude Bacon's enjoyment of aviation was considerably more sedate. She attended every possible flying meeting, joined the Aero Club, and spent much of her time at its flying ground at Sheppey or at Brooklands, meeting many of the young English aviators who were soon fulfilling a prophecy made to her by

Horace Short. 'England isn't behind the rest of the world at all,' he told her. 'In a very short time the world will have to own she is ahead.'

The pleasure Gertrude derived from being accepted by the flying fraternity was marred by the deaths of some of her new friends, such as Douglas Graham Gilmour, in flying accidents. Gilmour had taken her for her second flight, in a Blériot monoplane known as the 'Big Bat', which, with its passenger seat beside the pilot and windshield to screen her face, she had found considerably more comfortable but no less exhilarating than Sommer's biplane. At 1000 ft – only a year after many people had considered this beyond the capability of any engine – Gilmour had shut off the engine to demonstrate 'swooping', diving almost to the ground, a practice so dangerous that it later killed him.

In July 1912 Gertrude was invited for a flight in a small seaplane with which, much to the annoyance of some of the local residents, Edward Wakefield was conducting experiments on Lake Windermere. She was the first passenger, male or female, to be taken on the 42-minute circuit of the lake in the 'Water Hen', and the following day became the first woman to go up in a 'hydro-monoplane', a converted two-seater Depurdussin with a 70-hp engine capable of doing 70 mph: exciting but uncomfortable, with neither goggles nor windscreen. Two years later, she flew again in a seaplane, at Cowes, and found considerable improvements in comfort. It was only two months before the start of the First World War, but the seaplane was being sold to Germany, two Germans being also present at the trials. Gertrude was patriotically delighted to discover that the German navy crashed its acquisition only a few months after taking delivery.

She had her first experience of aerobatic flying at about the same time, at Hendon, with a seventeen-year-old pilot who took her up to 2000 ft and astonished her with the smoothness of the manoeuvre: 'The earth, somehow, had got into the sky . . . one couldn't fall out because the machine was the right way up all the time – only the earth, quite inexplicably, was up in the sky!' While the normal position of earth and sky were reversed, she was liberally spattered with oil from the Gnôme engine.

When war broke out Gertrude joined the Red Cross, and afterwards was kept busy giving lectures and generally boosting morale at camps, canteens, hospitals, convalescent homes and munitions factories. During the war, all non-military aviation stopped. When the British ban on civil aviation was lifted in May 1919, Handley Page Transport Ltd started a passenger service to Paris which opened in August. The first flight was a men-only press affair, but the following day Gertrude and the pilot's wife were promised places as the first women on the new air service. The £20 flight in an open twin-engined converted war plane, with no comfort for passengers other than slightly padded seats and the loan of fur coats, was cold and bumpy: it took four hours, a third of the time of the cheaper return by train and ferry.

Although a national newspaper had agreed to take an article 'First Woman Passenger Describes the New Service from the Feminine Point of View', when Gertrude submitted her copy she was told that the experiences of the first woman passenger had already been printed. That there had been no women on the inaugural flight was confirmed by the clerk at the Handley Page Transport office: 'If there was,' he told her, 'I can assure you she had a very well-developed moustache.' Gertrude was left only with the satisfaction of knowing she had achieved another first to add to the list of which she was so proud: first British woman to be a passenger on an airship, a biplane, a seaplane, and on a commercial flight between capital cities.

Gertrude Bacon married when she was fifty-five, was widowed five years later, had her first glider flight when she was over sixty and died in 1949 at the age of seventy-five.

3

The Good Old Crazy Days
in America

It was in France that a woman first became a pilot. Raymonde de Laroche, who called herself a baroness and claimed to be equally accomplished as an artist, a sculptress and an actress, was twenty-three when Charles Voisin offered to teach her to fly his single-seater box-kite aircraft. In March 1910 she gained the first licence issued to a woman pilot, and could see no reason why women should not fly as well as men: 'It does not rely so much on strength as on physical and mental co-ordination,' she claimed. The following year, two American women, Harriet Quimby and Matilde Moisant, followed in England by Mrs Hilda Hewlett and Mrs Cheridah de Beauvoir Stocks, received their licences.

The Americans were taught at the flying school set up by Matilde's brother, John Moisant, who, just over a year after Blériot conquered the English Channel, was the first pilot to fly across it with passengers: his mechanic and a tabby cat. He had started flying when, on a visit to France after escaping from a Central American revolution, he enrolled at Blériot's flying school and after a few lessons was convinced that flying was a better way of spending his life than running a hotel.

When he returned to the States John Moisant took with him some of the best European pilots. With three Frenchmen, an Italian, and two Americans, he formed the Moisant International Aviators and started the flying school at Hempstead, New York, where Matilde and her friend Harriet Quimby learned to fly. Even John's death in October 1910 during an aerial performance over New Orleans did not put the women off. Harriet, a New York journalist who claimed to come from a wealthy Californian family and to have been to private schools in America and Europe, but was probably the daughter of a Michigan farmer, was the first to gain her licence, in August 1911, followed less than a fortnight later by Matilde.

Both joined the Moisant exhibition team. In September 1911 Matilde won the Rodman-Wanamaker altitude trophy by reaching 1200 ft, then considered an amazing height. In October she upset the Nassau county sheriff by flying on a Sunday. When he tried to arrest her, she took off and flew to another airfield, and the judge to whom he applied for a warrant for her arrest ruled that flying on a Sunday was no more against the law than was driving a car. In November the two women went with the Moisant team to Mexico City, where they were the first women to be seen in the air. Harriet wrote an account of her Mexico trip for *Leslie's Weekly*, the New York publication for which she was drama critic and which, with the London *Daily Mirror*, sponsored her most famous exploit, a flight across the English Channel.

In April 1912, the pioneering British passenger E. Trehawke Davies crossed the Channel by air as a passenger, the first woman to do so, piloted by Gustav Hamel. Harriet Quimby reached England a few days later, and took off from the cliffs above Dover at 5.30 a.m. on Tuesday 16 April 1912. Hamel, who gave her some last-minute instruction about how to use a compass, was so certain that no woman could succeed in making the flight that he offered to wear her purple silk flying costume and make the crossing himself, landing at a deserted spot where she could be waiting to change clothes with him and take the credit. Naturally, she refused his chivalrous offer, but accepted the hot water bottle which he tied round her waist to protect her from the cold in the open cockpit.

Harriet landed on a flat sandy beach 25 miles south of her destination, Calais, the first cross-Channel flight by a woman pilot. She was treated as a heroine in Paris and London, as well as after her return to America with a new white two-seater Blériot monoplane. She reported on her flight for *Leslie's Weekly*:

> I was hardly out of sight of the cheering crowd before I hit a fog bank and found my needle of invaluable assistance. I could not see above, below or ahead. I ascended . . . hoping to escape the mist that enveloped me. It was bitter cold – the kind of cold that chills to the bones . . . A glance at my compass reassured me

that I was on my course. Failing to strike clear air, I determined to ascend again.

It was then that I came near a mishap. The machine tilted to a steep angle, causing the gasoline to flood and my engine to misfire. I figured on pancaking down so as to strike the water with the plane in a floating position. But, greatly to my relief, the gasoline quickly burned out and my engine resumed an even purr. A glance at the watch on my wrist reminded me that I should be near the French coast. Soon a gleaming strip of white sand flashed by, green grass caught my eye, and I knew that I was within my goal.

Three months later, Harriet Quimby, who had claimed never to have had an accident in the air because of her care in checking every wire and screw before mounting her machine, was killed at an air meet at Squantum Airfield near Boston. Her passenger for a flight round the Boston lighthouse was the organiser of the event, William Willard, whose son Charles was an exhibition pilot with the Curtiss team. He was a large man, and was warned that any sudden movement of his bulk in the Blériot would unbalance the aircraft. Neither he nor Harriet was wearing a safety-belt, and when the aircraft went into a sudden dive at about 1000 ft, first he and then the pilot herself were thrown out. No one ever knew the cause of the accident. They landed in shallow water and were both killed: ironically the Blériot came out of its dive and landed with only minimal damage.

Her friend's death, only eighteen months after her brother's, underlined the danger to which Matilde Moisant was constantly exposing herself. Her family put pressure on her to give up flying before she too was killed – she had had several minor accidents – and she agreed to make a flight in Texas her last. It was nearly more final than she had intended. As she landed in her Blériot, which bore her favourite number *Lucky Thirteen* as its name, it burst into flames: the fuel tank had developed a leak. The spectators were sure that the pilot would be cremated, but she crawled from the burning wreckage miraculously unharmed. Her heavy tweed flying suit, which with its knickerbocker divided skirt was considered almost as sensational as Harriet Quimby's purple silk, had saved her life.

Two other women – Ruth Law and Blanche Scott – were in the

air when Harriet Quimby was killed. Blanche had made her first flight in 1910 with Charles Willard, soon after driving a car from New York to San Francisco to prove that long-distance motoring was so easy that even a woman could do it, and proceeded to prove that women could be as daring as men in the air. Glenn Curtiss, who disapproved of women trying to fly, reluctantly accepted her as a pupil, but put a throttle block on the aircraft to prevent it leaving the ground. On 2 September 1910, in spite of the block and the instruction that she was not to attempt to take off, Blanche rose 40 ft and became the first American woman to make even such a short solo flight. Curtiss grudgingly allowed her to carry on, and even took her into his exhibition team, with which she made her public début at an air meeting in Chicago at the beginning of October.

For several years, Blanche Scott flew for various exhibition teams, satisfying the public's wish for excitement by flying upside down 20 ft from the ground, under bridges, and specialising in a spectacular 'Death Dive' in which she plummeted from 4000 ft and levelled out only 200 ft from the ground. In 1916, at the age of twenty-seven, she retired. 'In aviation there seems no place for the woman engineer, mechanic or flier,' she explained. 'Too often, people paid money to see me risk my neck, more as a freak – a woman freak pilot – than as a skilled flier. No more!'

It was not Blanche Scott, but Bessica Raiche, who was officially honoured as the first American woman aviator, although she did not leave the ground until a fortnight after Blanche's disobedient hop, which was officially ruled to have been accidental. Bessica's flight was intentional. Her interest in aviation had been aroused while she was studying music in Paris by the exploits of Raymonde de Laroche. She returned with a French husband, with whom she built a flimsy aircraft of bamboo, silk and wire in their living room in Mineola, New York. Bessica had never flown, nor indeed had any flying tuition, when she made her first solo flight in it.

After a minor mishap on her fifth flight, she decided that her long skirts were a hazard, and took to flying in riding breeches. She was already considered eccentric for wearing bloomers for sport and for indulging in such unfeminine activities as shooting, swimming and driving, as well as the more acceptably ladylike pastimes of music and painting. The Raiches developed a small

cottage industry with two more of their silk and bamboo constructions, and then formed a French-American Aeroplane Company to develop the business, using light piano wire to improve
performance by reducing the weight of their aircraft. When
Bessica's health forced her to give up flying the Raiches moved to
California, where, still determined to enter a man's world, she
became a doctor.

The flying career of Ruth Law, who was on her first flight when
she saw Harriet Quimby fall to her death, was longer. Her
brother, Rodman Law, was a steeplejack, and as the 'Human Fly'
his antics included being shot from a cannon from the top of a
New York skyscraper; he had made his first parachute jump from
the Statue of Liberty. Ruth, who made daily exhibition flights and
took passengers for joy rides for a Florida hotel throughout her
first winter as a licensed pilot, was the first woman to fly at night,
making a twenty-minute moonlit flight round Staten Island in
November 1913. Three years later, when the Statue of Liberty
was first spotlit, she gave a spectacular night-flying display with
magnesium flares blazing from her wingtips and LIBERTY in
illuminated letters on the bottom of her aircraft.

Acclaimed as the first woman to loop the loop, Ruth Law took
her flying seriously, and in 1916 set a women's altitude record of
11,200 ft. She was not content with mere women's records and was
furious that she was nearly 4000 ft short of that set by a male pilot.
Later in the year, in a 590-mile non-stop flight from Chicago to
New York, she broke the American cross-country record in an
open Curtiss pusher, with auxiliary fuel tanks increasing the capacity from 8 to 53 gallons, and a small shield round her feet to protect
them from the November cold. Her cross-country navigation was
helped by her invention of a map roll tied to her knee, so that she
could turn a knob to wind the 8-in. strips she had pasted on to cloth
and wound round a roller without leaving go of the controls.

Her tumultuous reception in New York brought Ruth Law
fame and an income of up to $9000 a week for her later exhibition
flights. Even this did not persuade the authorities that a woman
might reasonably fly during the First World War with the United
States Army. She was so incensed by her rejection that she wrote
an article for the magazine *Air Travel:* 'It would seem that a
woman's success in any particular line would prove her fitness
for that work, without regard to theories to the contrary.' As the

first woman to wear an army NCO's uniform, she was nevertheless allowed to raise money for the Red Cross and Liberty Loan
through exhibition flights, during one of which, in September
1917, she set a new women's altitude record.

After the war Ruth Law toured Japan, China and the Philippines with her aerial exhibition, and with her husband ran a
three-plane flying troupe called Ruth Law's Flying Circus which
flew in tight formation with Ruth in the centre aircraft. Her solo
stunts included racing low round a track in a speed race against a
car, and car-to-plane transfers, during one of which the stunt girl,
who was to leap from the car to the rope ladder dangling from the
aircraft, missed and was killed when she hit the ground at 60
mph. In her most dangerous stunt, Ruth climbed out of the
cockpit to stand upright in the centre of a biplane's wing while the
pilot looped three consecutive loops. Her husband was so
alarmed at this unnecessary risk – they did not need the money –
that he announced her retirement in the press. Ruth obediently
retired, explaining that 'Things are so proper now . . . so many
rules and regulations . . . The good old crazy days of flying are
gone.'

Ruth Law was not the only well-known American woman pilot
to be refused permission to fly in the war. The Stinson sisters,
Katherine and Marjorie, were allowed to train Canadian pilots for
flying service in Britain, to give international exhibitions, to make
and break records and to raise money for the war effort by flying:
but military service was strictly for men only. They had entered
aviation almost accidentally, when Katherine, who was sixteen
and weighed only just over seven stone, won a balloon trip in a
raffle and hit on the idea of financing her music studies by
becoming a stunt pilot. As she had been assured that earning
$1000 a week was not impossible, she sold the piano she had won
in a talent contest for $200 and borrowed another $300 from her
father to pay for flying lessons. During her first lesson she was
convinced that the aircraft was out of control when the pilot
banked, but she wanted to go up again immediately. Although
her instructor was convinced that she would either be killed or
catch pneumonia, and refused to continue her lessons, she
persuaded the Swedish pilot Maximilian Theodore Liljestrand,
commonly known as Max Lillie, to take her on as a pupil at his
new flying school.

Less than a fortnight after Harriet Quimby's death, Katherine was given her licence, the fourth awarded to an American woman. Her training had been eventful: she and Max Lillie had been arrested for landing for a yacht club luncheon in a public park, and on her first solo flight she had had to make an emergency landing when her engine failed. Because she was so young, she decided to wait a year before starting her exhibition career, and delayed her début until July 1913 before touring to flying meetings all over the country.

In April 1913, Katherine Stinson and her mother Emma formed the Stinson Aviation Company 'to manufacture, sell, rent, and otherwise engage in the aircraft trade', with $3070 capital borrowed from friends. The following month, Katherine bought a $2000 Wright 'B' which had been modified by Lillie. To the amusement of the mechanics at Cicero Field she insisted on washing the accumulation of grime from her new acquisition, and found several worn fittings. After this, she always insisted on her aircraft being kept spotless as a safety measure: when Max Lillie's aircraft broke up and he was killed later in the year during an exhibition flight, she blamed the accident on his carelessness about maintenance.

In September 1913 she was flying at the fairground at Helena, Montana, when an airmail route from the fairground to the centre of Helena was sanctioned, and for four days she carried mail by air, the first woman to do so.

Marjorie Stinson soon followed her sister into the air, although at first Katherine tried to dissuade her because of the danger. Neither would teach their brother Eddie to fly, although they taught Jack: Eddie, they considered, was too wild and drank too much, but by working as their mechanic for the summer of 1915 he nevertheless saved enough for lessons at a flying school in Ohio, and when the Stinson Aviation Company had to move because the embryonic American Air Service requisitioned the parade ground from which they had been operating, it was Eddie who found and cleared a new 750-acre site. On their private airfield at San Antonio in Texas, the Stinsons opened a flying school, with Marjorie as chief instructor and Katherine's exhibition flying providing the financial backing.

Like Ruth Law, Katherine Stinson was acclaimed as the first

woman to loop the loop, and had developed a 'dippy twist' loop in which at the top of a vertical bank the aircraft rolled wing over wing. In November 1915 she made eighty consecutive loops, flying upside down for thirty seconds and executing a series of spins. In December, determined to out-do a male pilot, Art Smith, who had looped the loop at night, she added magnesium flares to her aircraft and traced the letters CAL in the night sky, then looped, flew upside down, and spiralled to 100 ft off the ground, trailing showers of sparks. For the first six months of 1917, she toured China and Japan, where no woman had flown before.

Twenty-five thousand people watched her first performance in Tokyo. Bonfires lit the sky as she performed fifteen minutes of night stunting, tracing the letter S in the sky with fireworks. Her reception was enthusiastic: Japanese women hailed her as a female emancipator, and Stinson youth clubs were started. A Tokyo schoolboy wrote to her that he wanted 'to make myself air-man' and praised her skill to the skies: '. . . when I saw that you were flying high up in the darkest sky I could not help to cry: you are indeed Air Queen!' In Peking she performed for Chinese dignitaries on the sacred ground in front of the Temple of Agriculture. During a spiral descent after three successive loops, etching a path in the sky with a trail of smoke, the rudder bar dropped off: she could keep directional control only by bending over to manipulate the stub by hand, flying blind as she did so.

Her tour was cut short when America entered the war in Europe: she volunteered for army flying service, but when she was turned down she made a 640-mile fund-raising flight from Albany to Washington, DC, in a Curtiss military trainer. Later in the year she established and then broke endurance and distance records across America. During the war, the Stinson Flying School trained Canadian pilots who then joined the British flying services: there was no flight training available in Canada. Marjorie and Katherine instructed, and Mrs Emma Stinson was their business manager. They had two male mechanics, one of them Eddie, who was constantly at odds with his mother: she was strict and refused to allow gossip, while he was irrepressibly chatty and fun-loving.

In November 1915 Marjorie, who was known as the 'Flying

Schoolmarm', had a class of five Canadians who, after her tuition, all joined the Royal Navy Flying Service. In January 1916, the Stinsons' school had fourteen pupils, mostly Canadians, and four serviceable aircraft. By the end of March, there were twenty-four students, and a new engineer called Brock who updated their Wright trainer and built identical Gnôme-powered 'Brock loopers' for Katherine and Marjorie.

By August 1917 only two of the trainer aircraft were left, a Wright 'B' and an Ox-Burgess made up from three old engineless Burgess biplanes and an 80-hp Ox engine which Mrs Stinson had bought for a total of $350. When the American government banned civilian flying, the Stinson school was liquidated and Marjorie became a draughtswoman with the Aeronautical Division of the US Navy in Washington, DC.

Katherine drove ambulances in England and France and did some flying for the Red Cross. She married a lawyer, Mike Otero, in 1918, and in the same year toured in Canada. When she landed unannounced at Camp Hughes in Manitoba, she was arrested, but managed to persuade her captors that she had come not to spy, but to entertain, and was allowed to give a performance for 10,000 troops. Her stunts included nose-diving so steeply that often the onlookers thought she had actually crashed. She did have one crash, with a broken piston, and landed in a wheatfield, where she was asked by a surprised farmer: 'Is that one of them newfangled threshing machines?' At Calgary, Alberta, the army officers were particularly impressed by her 'bombing' exhibitions, and made her their special guest.

She left Calgary after a week's engagement with a sack of 259 specially stamped letters to inaugurate an airmail service to Edmonton where, after a spectacular aerial display during a 'Katherine Stinson Night', she clipped a parked car as she came in to land, fortunately causing no serious damage either to car or aircraft: the headlights of the motorists had blinded her. A few weeks later, she asked to join the United States Aerial Mail Service as a pilot. Permission was granted only after some argument among the officials, and she then insisted on having twin joy-sticks fitted to the airmail aircraft: she was used to the Wright arrangement, and refused to fly with the single joy-stick and rudder bar controls which had become standard. After one delivery flight with an escort between New York and Washington, which the press treated as a race, she retired from the airmail service, which she had fought so hard to join, with no explanation.

In 1920, because of ill-health – exhaustion and tuberculosis – Katherine Stinson gave up flying. Eddie Stinson was turned down for US Air Service flying because of his weak lungs and heavy drinking. He became an army instructor and stunt pilot, popping light bulbs with a wing tip and dropping crates of eggs without breakages to advertise the safety of delivering mail by parachute. Among the passengers he took up for joy-rides was a milliner's model called Estelle Judy, whom he married. On one occasion, she took over the controls in the air while Eddie climbed on to a wing and leant out so alarmingly far to change a plug that his mechanic fainted.

Eddie Stinson's career was alternately drunken and brilliant. In 1920 he and Jack joined forces to form the Stinson Aeroplane Company for the manufacture of aircraft in Dayton, Ohio. The Stinson Airplane Syndicate was formed in 1925, and the Stinson Detroiter was Eddie's brainchild, although he had considerable difficulty obtaining funds to develop it as he was by then Dayton's most notorious drunk. His mother, who had never made her peace with Eddie although he supported her in some comfort in a house in the suburbs, became increasingly paranoid, calling at her son's office dressed in rags to demand money. Eventually she was committed to a mental hospital.

During the 1920s aerial entertainment spread through America as flying gipsies toured with spectacularly death-defying displays and touted for joy-riding custom. Towards the end of the First World War American aviation production had been stepped up to keep pace with the demand for military aircraft, with the result that when the war ended there was a glut of war-surplus biplanes. A single-engined Curtiss JN-4 biplane, popularly known as a 'Jenny', could be bought in its crate for $600. After the war the 'Jenny' was used for postal services; many of the men who had flown in the war, or who had received pilot training too late for active service, bought 'crates' which needed no more than a field to land in. They flew from farm to farm, landing in fields where a barn could provide some shelter at night – hence the name 'barnstormers' – and performing aerobatics to attract customers. In October 1919 the editor of the American magazine *Aviation* wrote:

One of the most interesting phases of present aviation activities is the great number of small companies engaged in exhibition flights and in passenger flights of short duration. Such work has not yet reached the dignity of aerial transportation work, but nevertheless both activities have considerable educational value for the general public.

The education of the general public, although usually only of secondary importance for the pilots, whose chief aim was to find a way of supporting themselves by flying, proceeded wherever one or more pilots landed to offer short flights. Many people had their first experience of flying through a joy-ride with a barnstormer, and many pilots who later achieved a more serious position in aviation started as barnstormers, among them Charles Lindbergh, the first person to fly solo across the Atlantic. Some added a deliberate element of comedy to their displays, like William Kopia – 'Wild Bill' – of Newark, who appeared half-way through air shows dressed as a female opera star. The star would buy a ticket for a joy-ride, with suitable announcements over the loudspeaker, and as she climbed into the passenger seat of a 'Jenny' the pilot would climb out and run to the hangar for something he had forgotten. The opera star would then accidentally hit the throttle; the aircraft would lurch forwards, and to the delighted alarm of the crowd would suddenly leap into the air and stagger back – at which point the star would revert to his true role as a skilled aerobatic pilot.

Women who wanted to make a career in aviation found the only way to make it pay was by barnstorming, and some with a less solemn intention became well known for their antics as parachutists and wing-walkers. Helen Lach, who left a job as a waitress to become a parachute jumper at aerial exhibitions, explained that she got 'a wonderful thrill' as she sailed down to earth from the clouds. Lillian Boyer was another waitress who took to the air: in 1921, on her second flight, she climbed out of the cockpit on to the wing, and later in the year made her first plane-to-plane mid-air change. After five months of intensive training, she toured America and Canada with wing-walking stunts, car-to-plane changes, and parachute jumps, performing altogether in more than 350 shows and retiring only in 1929 when regulations on low flying forced many barnstormers out of business.

One woman who financed her early flying by parachute spectaculars – later gaining a reputation and a career in serious aviation – was Phoebe Fairgrave. She made her first jump when she was seventeen, and toured with the Glenn Messer Flying Circus, making a record jump from over 15,000 ft in July 1921 and developing a double parachute jump in which she cut the first parachute loose to free-fall before releasing the second. By the time she was twenty she had her own flying circus, and married Vernon Omlie, her chief pilot and the man who had taught her to fly. Together, they ran the Phoebe Fairgrave Flying School until they earned enough to start their own flying school, Mid-South Airways, in Memphis.

Ethel Dare, the 'Flying Witch' – whose real name was, more prosaically, Margie Hobbs – was a circus trapeze artist before she took to wing-walking in her teens in 1921. One of her stunts was a plane-to-plane transfer on a rope ladder suspended beneath one of the aircraft, but when a male pilot was killed doing the same stunt, it was banned and she went back to land-based circus work.

The first black woman to gain a pilot's licence, Bessie Coleman – who because of her colour had had to go to France to learn to fly – earned a living barnstorming while she saved towards her own flying school, but was killed when she was thrown out of an aircraft which she was testing: she had saved nearly enough to achieve her ambition. Gladys Roy, who used to dance the Charleston on the upper wing of a biplane, died when she walked into a spinning propeller on the ground.

The self-styled 'Greatest Aerialist and Transfer Artist' and 'Queen of the Air' was Mabel Cody, whose Flying Circus, complete with three Sky Rings, was one of the most popular on the aerial entertainment circuit. Not content with mere wing-walking, she used to dangle from one of the aircraft's struts, transfer from one aircraft to another, and make high-speed transfers by rope ladder from car to aircraft. Her most spectacular stunt was falling off the wing, to do an 'Iron Jaw Spin' in the slipstream.

Aerial circus acts and barnstorming may sound like the lighter side of aviation, but any circus pilot or barnstormer who survived, both physically and financially, had proven flying ability, combined with unusually quick reactions and undoubted stamina. But, by the 1930s, the good old crazy days of flying were fading, even in America. Nevertheless American dare-devilry had provided immense pleasure for many, both participants and onlookers, and an income for men and women who could not otherwise have afforded to fly.

4

Flying Aristocrats
and their Aerial Motor Cars

In the late 1920s and early 1930s, several titled British ladies took up flying. Two had more money than luck, or perhaps sense, to the benefit of the national exchequer, as they died intestate with their hired pilots over the Atlantic: in 1927, the state inherited the fortune of the sixty-three-year-old British-born Princess Anne of Loewenstein-Wertheim, and in 1928, the Hon. Elsie Mackay, daughter of Lord Inchcape, left £500,000 which her father presented to the government towards the repayment of the national debt.

Those who survived had the good sense not to attempt the Atlantic, but inflicted on themselves long-distance ordeals across and even between continents. They were as notable for their eccentricity as for their stamina, and shared a firm belief in the future of air travel, although the time they spent waiting on the ground for repairs whenever they crashed their flimsy aircraft was often more than they spent in the air. The aggressively feminist Lady Heath and her gentler friend and rival, Lady Bailey, passed each other as they flew in opposite directions across Africa in 1928. Two years later, the Hon. Mrs Victor Bruce decided, on the spur of the moment, to buy an aircraft and fly round the world, surviving innumerable mishaps, while, in her sixties, the deaf and unsociable Duchess of Bedford used a private chauffeur-flown aircraft as an aerial motor car.

LADY HEATH

Lady Heath, who was born Sophie Mary Pierce Evans in County Limerick in 1896, was a militant advocate of both women's rights and aviation. By the time she embarked on a flying career at the age of twenty-two, she possessed a Dublin science degree and had lectured at Aberdeen University. As soon as she gained her pilot's licence in 1925, as the inaugural pupil at the London Aeroplane Club, she started to campaign for the advance of civil aviation in general and for equality for women in the air in particular. She did not yet have the advantage of wealth: until she acquired her second elderly husband, the widowed Mrs Sophie Elliott-Lynn had to earn her own living, which she was determined to do as a commercial pilot.

In 1924 the International Commission for Air Navigation had specifically excluded women from 'any employment in the operating crew of aircraft engaged in public transport', and had made the first requirement of physical fitness for an aspiring commercial pilot that 'he must be of the male sex'. Mrs Elliott-Lynn's first move in her campaign for equal rights in the air was to write to the international commission stating her credentials, which she felt gave her a greater insight into women than had been available to the male medical sub-commission responsible for the ban: she had a degree in physiology, was a champion high-jumper as well as a pilot, had founded the Women's Amateur Athletic Association of Great Britain, and was the only woman member at the Olympic Congress in Prague in June 1925 to give evidence on the advisability of allowing women into world athletics.

She was subsequently summoned to prove that she could function competently as a pilot whatever the time of the month, and demonstrated that even her landings, the most skilled part of a pilot's work, were unaffected by the disability of being a woman. While she waited for her campaign to take effect, she qualified theoretically for her B licence.

The international system of licensing was drawn up in 1919 and brought into effect in Britain with an Air Navigation Act the following year. The A licence, commonly known as a PPL – Private Pilot's Licence – was the lowest grade, entitling the holder to fly for pleasure and to carry passengers and goods, but not to charge for doing so. It could initially be passed at the minimum age of seventeen after only three hours' solo flying, a medical examination and tests of flying ability and technical knowledge. It was considerably more difficult to become a professional pilot entitled to fly for hire or reward: the commercial qualification was the B licence, for which applicants had to be between nineteen and forty-five, physically fit, and have flown solo for at least one hundred hours. The technical knowledge required covered mechanical theory, meteorology and navigation, and the practical tests included several cross-country flights of over two hundred miles, as well as altitude and night flying. An A licence ran for a year, and could be renewed for 5s. and three hours' solo flying in that time. The initial cost of a B licence was thirteen guineas, and it had to be reviewed every six months: the pilot must have flown six hours' solo since his last licence was issued, be medically re-examined and pay a fee of 15s. 6d. In addition, there were graded navigators' and ground engineers' licences.

As Mrs Elliott-Lynn still could not take paying passengers, she at first earned her living in the ranks of exhibition and stunt pilots and by flying newspapers between Paris and London. In May 1926 the earlier unanimous ban on passenger-carrying women pilots was unanimously reversed, with the proviso that women should be medically re-examined every three months, while men were considered capable of remaining fit for six months at a time.

Mrs Elliott-Lynn immediately set about proving that women could fly as well as men. In May 1927, with Lady Bailey as passenger, she set an altitude record of 16,000 ft, acknowledged as a feminist achievement in *Flight Magazine:* 'It is still a common practice for women, as aviators, to be rather disdained. Mrs Elliott-Lynn has perhaps done more for her sex than any other woman.' She competed successfully in several races, and in July made a 13½-hour tour of 79 British aerodromes, flying 1300 miles in a day. The following week, she enlightened 8000 people in Ireland about the history and progress of aviation, and in September had an even larger audience for a similar lecture at Aberdeen. Between Ireland and Scotland, she fitted in a tour of central Europe in a new Avro Avian. In October Mrs Elliott-Lynn relinquished the name of her deceased husband when she married the elderly and wealthy Sir James Heath: in November, Sir James and Lady Heath left for Cape Town by ship, with one of her four aircraft, an Avro Avian, in a crate.

Lady Heath planned to create a record, and attract as much publicity as possible, by a solo flight from Cape Town to London. Although three survey parties had investigated Africa's flying potential soon after the end of the First World War, much of the continent had still not been explored by air. The first solo flight from England to Cape Town was made by Flight Lieutenant R.

Bentley, an officer in the South African Air Force, who reached Cape Town in a de Havilland Gipsy Moth three months before Lady Heath arrived by ship. He had taken 27 days: Lady Heath's flight in the opposite direction took four months, although much of the time was spent on the ground. Bentley sold Lady Heath his maps, the only set she could find in Cape Town for her proposed route, but then decided he would fly to England with his new wife, took them back, and lost them. For much of her navigation she had to rely on small-scale route maps and tracings or torn-out pages from an atlas.

By the time she reached Johannesburg, she had flown 70 hours, many of them by giving joy-rides which earned £1200 for South African flying clubs and to set up flying scholarships in England; but she had missed a civic reception and tribal war dance laid on in her honour when she lost her way and landed at the wrong airfield.

Lady Heath travelled equipped for almost any eventuality. As well as the clothes she wore for flying, basic medical supplies of morphine, quinine, iodine, vaseline and a couple of bandages, she took a Bible and some novels, six pairs of silk stockings, two silk day dresses and an evening dress, two blouses, a jersey and a white flannel skirt, black satin shoes and tennis shoes, and a tennis racket, but only one change of underclothing. The black silk evening dress and high-heeled shoes were worn on several social occasions – she was entertained at various embassies and government houses, even using her tennis racket – and, more bizarrely, in the Libyan desert. While she was supervising repairs to her aircraft, she was visited daily by an armoured carload of Italians, and did not wish to be outshone by their immaculate white gloves.

To protect her neck from the sun, she wore a fur coat while she was flying, even when the temperature in the cockpit rose above 120°F. Nevertheless she was stricken violently and suddenly by sunstroke after flying for six hours from Pretoria over the meanderings of the Limpopo River and above the Matopos hills: somehow she landed without injuring herself or doing any more than minor damage to her aircraft, and was then unconscious for four hours. It took her a weekend to recover enough to continue to Livingstone. There she found Bentley and his wife Dorys, who suggested that they should fly on in company. Lady Heath accepted grudgingly, as she wished to be the sole centre of attention for her achievement, but she was grateful enough when she discovered that she was not allowed to fly without a chaperone over the Sudan, where a district commissioner had recently been killed. Bentley accepted £5 per flying hour to act as her escort to Khartoum, where he gallantly turned back to chaperone Lady Bailey in the opposite direction.

In Cairo, Lady Heath smugly attributed the success of the flight so far to the care she had given her engine: 'I did the tappet clearances every day, no matter how short the flight was, and cleaned the petrol and oil filters. Only once did I fly in the heat of the day, and I never flew at less than 7000 ft to get the cool air. I ran my engine at 1700 rpm throughout and did one to three hours' routine work daily.' There had been setbacks, and although she had flown 72 hours in 16 days, she had spent nearly as many days on the ground. She had been robbed during the night in Nairobi, where she was depressed by the lack of favourable publicity her efforts had attracted. The implication of aviation officials that she had been slack in letting them know her movements and so had made it impossible for them to keep up with her 'vagaries' struck her as most unfair: she had sent numerous telegrams, which she claimed had either not been delivered or had been garbled, and said that 'These disgraceful postal facilities constituted the main danger of the trip.'

As if to give the lie to this claim, when she left Nairobi she had considerable difficulty in lifting her aircraft, with its full load of petrol and her 112 lb of equipment, off the ground, and narrowly cleared a 10,000 ft escarpment only after flying backwards and forwards for half an hour to gain height, making a 50-mile detour following a railway line and throwing out her tennis racket, a pair of shoes and some books: 'one of the most exciting moments of the trip . . .'

Bentley came to her rescue again as escort before she was allowed to cross the Mediterranean, although Lady Bailey had been allowed to cross from north to south unescorted: but Africa was easier to find than Malta. Although when she had left a dance at the British Residency in Cairo to send a telegram to Mussolini asking for assistance, he had cabled back, 'Have put a seaplane at your disposal', she waited in vain: it had come down with engine trouble and drifted for four days on a rough sea 40 miles from the coast.

Lady Heath was unaware that she had been shot at over the North African coast until a bullet hole was discovered in one of the wings of her aircraft in Tunis. Her greatest fear was of crossing the sea, for which she wore two inflated motor tyres as a lifejacket: at 7000 ft they burst and she was left with shreds of rubber hanging round her neck. She nevertheless managed to arrive in Rome after an eight-hour flight looking, according to the aviation writer Stella Murray, 'as if she had stepped out of a bandbox, having changed her flying helmet for a little black cloche straw hat'. Appearances were important to Lady Heath, who managed to change in mid-air so that she could emerge from her aircraft looking her best, and claimed that flying was so safe that a woman could 'fly across Africa wearing a Parisian frock and keeping her nose powdered all the way'.

LADY BAILEY

Lady Mary Bailey had a more relaxed attitude. When the two ladies were joint guests of honour at a reception in Khartoum, Lady Heath appeared in evening dress and Lady Bailey in a tweed flying suit. The only equipment she considered essential as she pottered down to Cape Town and back was mosquito boots – on the return she was delayed for a month at Leopoldville by infected mosquito bites – although tinted goggles and a sun helmet were 'desirable'. Lady Bailey's approach to flying was equally relaxed, although in spite of her apparent vagueness, her modesty and her off-hand way of talking about flying in the simplest and most amateur manner, it was she and not her more aggressive friend Lady Heath who, in 1929, was the first woman to receive the Britannia Trophy for the year's most outstanding air performance.

The daughter of an Irish peer, wife of a South African millionaire, and mother of five children, Lady Bailey gained her A licence a year after Lady Heath, and then made the first female flight across the Irish Sea. She went on to study navigation and become the first woman to be given a blind-flying certificate. There was little to choose between the two ladies when it came to flying ability: Lady Bailey bettered the altitude record set by Lady

Heath, who then raised it again, and in races sometimes one, sometimes the other, was in the lead. Lady Bailey's decision to fly to South Africa did not seem to her extraordinary: 'Having learnt to fly and having acquired a British light aeroplane, able with reasonable luck to take its owner anywhere in the Empire, it seemed a natural thing to use it to join my husband in South Africa.'

After flying 8000 miles, she apologised to Sir Abe Bailey for arriving late at the Cape because of 'getting muddled in the mountains': it had taken her 28 flying days, but delays had more than doubled the time. In Tanganyika, she had turned her aircraft upside down, disgusted with herself because 'through insufficient care and a lack of knowledge of how to handle a machine in landing at an altitude in the heat of the day' she had damaged it irreparably. For a less wealthy pilot, this might have meant the end of the adventure: Lady Bailey was, however, in the fortunate position of being able to cable her husband for a new aircraft, and merely had a nine-day wait for it to arrive.

She had not been in South Africa long when she decided that, having flown out, she might as well fly back. After a two-month shooting trip in the Transvaal, she set out on an untried route across central Africa and the Sahara. Several weeks went by when she crashed near the start of her venture and had to wait for repairs, and although she set out from Cape Town in May it was September before she left Rhodesia. The lack of adequate maps caused her many 'anxious moments', and she resorted to following dried-up river beds between mountains partially obscured by mist. On her return to England, ten months after she had set out, her flight from London to Cape Town and back was hailed in the press as the greatest solo effort ever accomplished.

THE HON. MRS VICTOR BRUCE

As the Hon. Mrs Victor Bruce had been flying for only a few weeks when she set out round the world in 1930, she wisely cheated over the Atlantic and Pacific oceans by taking her aircraft on a ship. Even before her marriage to a racing driver, Mary Petrie had always been adventurous. As a child, she was delighted when her governess's cart overturned or her pony bolted, and as a teenager she roared around the countryside on a motorbike, with her dog in the sidecar. When she acquired an Enfield Allday sports car, she found herself repeatedly in court for speeding.

In the late 1920s, Mary Bruce, sometimes with her husband and sometimes alone, indulged her love of speed by competing in Monte Carlo rallies and setting endurance and speed records both on land and, in a speedboat, at sea. She had no intention of taking up flying until one day when she was window-shopping in London she saw a Blackburn Bluebird with a ticket on it saying that it was 'ready to go anywhere'. She asked the price: '£550 . . . and chromium plating is £5 extra.'

Intrigued, she went home, took out an atlas and traced a line round the world: across the English Channel to Belgium, Germany, Constantinople, across Turkey to Syria, along the Euphrates to Baghdad; then to Basra, the Persian Gulf, across India to Rangoon, over the Burmese and Siamese jungles to French Indo-China, Hong Kong, Shanghai, across the inch of the Yellow Sea to Tokyo. She paused – the Pacific looked so large that she would cross it by ship – but she resumed her line at Vancouver, down the west coast of America to San Francisco, and on to New Albany to find the house her American mother had left more than forty years earlier. This, she decided, was her destination, but then drew the line to New York, where she would again cheat by steamer: even the atlas could not minimise the danger of the Atlantic for someone as inexperienced as she would still inevitably be. She would not, however, disembark in England, but would continue by ship to France so that she could fly back triumphantly to England.

Having decided on a route, she bought the aircraft and visited the Air Ministry, which in due course provided sixty-six pages of weather information. She then proceeded to the AA, which at the time supplied air route maps: 'When did you learn to fly?' she was asked. 'I haven't yet, but I will before I go,' she promised. She had soon learnt both to fly and to navigate, getting up at seven to study, and in two weeks she had her licence.

Although she refused to learn Morse, she agreed to take a primitive wireless set which gave her more ground contact than had been available to Lady Heath or Lady Bailey: it could not receive any messages, but she could put a plug into the relevant socket to send for help or impart general set information while she was in the air. When it came to a choice between a dictaphone and a parachute – there was not enough space for both – she chose the dictaphone, so that she could send progress reports home. Apart from the dictaphone and essential remedial equipment for the aircraft and engine, she took only a small shoulder bag containing her husband's greatly prized pocket compass, her passport, log book, and a water bottle, a sun helmet, her flying clothes and two light cotton frocks as well as, like Lady Heath, an evening dress.

The Bluebird bore the registration G-ABDS – 'a bloody daft stunt', according to those who did not expect her to reach even the far side of the English Channel. When she said good-bye to her husband and young son, 'a terrible loneliness' came over her and for a moment even she doubted her ability to accomplish the task before her. It took her five months to circumnavigate the world, with frequent crashes and adventures. Many of her accidents could be put down to inexperience, although her expertise in motor racing was undoubtedly a help. Her first brush with death came as she was following a railway line after leaving Belgrade, to find as she flew round a sharp bend that it disappeared into a tunnel: at 100 mph she just managed to fling the aircraft round between the mountains. Over Turkey, she carelessly knocked the rudder and sent the aircraft into a spin only 500 ft above the ground: she had been trying to reach out to wipe oil from the windscreen, and was only just able to regain control in time.

Some of her landings were distinctly unorthodox. When she could not find the aerodrome at Ankara, she used the stadium, from which she first cleared a crowd by dropping smoke bombs. Another emergency landing in Turkey led to a meeting with 'le Turc Vénérable', reputedly the oldest man in the world: his touch was supposed to bring luck, which Mrs Bruce certainly needed in spite of her claim that she was 'just an ordinary aviatrix'. After yet another unscheduled landing, in the Syrian desert, it was announced in the press that she had been abducted by a sheikh. The truth was scarcely less romantic: her abductor, an officer on an Arab horse, had galloped at the head of a troupe of horsemen in red fezzes to rescue her from an importunate Armenian.

When a fall in oil pressure forced her down on quicksand beside the Persian Gulf, she plugged into 'Flying in trouble', but no one heard. To the sound of splintering wood and the smell of escaping petrol, she found herself hanging by her straps, the tail of the aircraft up in the air and its engine buried in sand. Some Baluchi tribesmen gave her what assistance they could as she attempted to repair the damage with a small knife and clean the engine with her toothbrush – she had lost her toolkit – and their chief kept guard during the night as she tried to sleep on the sand.

She was soon weak from hunger and thirst which the flyblown dates and brackish water brought by the Baluchis did little to alleviate, and a 'perfectly good message' she had made for her husband on a dictaphone record was ruined when they mistook it for chocolate. When they discovered that she was a woman, they gave her a big scarf to cover her bare legs and shorts: as it kept tripping her up, she changed into pyjamas, which seemed to satisfy their desire for maidenly modesty. On her second evening in the desert, she had three unwelcome visitors, who eventually left with her money and coat: one climbed into the aircraft and played with the instruments, one tried to put on her evening dress, and one merely sat and leered lasciviously. The next morning, she decided she had had enough and started to walk the 50 miles to Jask, but collapsed exhausted at the first village she reached.

When news of her crash reached Jask – one of the Baluchis had taken 48 hours and swum several shark-infested creeks to deliver a message – a British rescue party set out at night by sailing boat. In the morning they found first the aircraft, then the pilot's shoes, and at last the pilot. In Jask, where the temperature was 100°F in the shade and she was suffering from dysentery after her three-day ordeal in the desert, Mrs Bruce's 'great nerve and endurance' were duly admired. At last the aircraft continued its eventful journey, struggling through a sandstorm in the Sind desert, attacked by vultures, and suffering further damage when it rolled over on landing at Calcutta. Tropical rainstorms, an emergency landing and hazardous take-off in the jungle, malaria, elephant riding and big game hunting added further variety both on the ground and in the air.

In Laos, a Siamese Roman Catholic priest gave Mrs Bruce a St Christopher, which she was convinced came to her rescue as she flew through cloud over the mountains and emerged over the Indo-Chinese mainland without once seeing the sea. It was her first experience of flying above cloud, but she had been warned of the danger of hitting a mountain if she started to descend too soon. With only an hour's fuel left, feeling frightened and lonely, she dictated a message: 'I'm lost above the clouds. This will be my end. I've done the best I can, and if I come through it will only be by the grace of God. Good-bye.' Then she dived through the cloud, to see miles of swamp, and beyond the swamp a railway – and Hanoi, where she was greeted with bouquets and granted the 'Order of the Million Elephants and the White Umbrella'.

The governor of Foochow replied to her advance request to land on her way to Shanghai with a telegram: 'Sorry impossible for you to land; too busy with the war'. As she flew over the fighting, after a night start to be sure of completing the 650 miles to Shanghai – only 50 miles short of the Bluebird's range – she saw Chinese soldiers dashing about beneath her, and added to the confusion by dropping an impartial smoke-bomb. There was another enforced change in route when she was forbidden to fly over the Japanese five-day Grand Manoeuvres attended by His Imperial Majesty the Japanese Emperor: no one was allowed to look down on an Emperor of Japan.

The diversion to Korea made the Hon. Mrs Victor Bruce the first person to fly direct from Shanghai to Korea over the Yellow Sea, a flight which, she was told, was the longest solo sea crossing since Lindbergh's transatlantic flight two years before. When she was allowed to proceed to Osaka and Tokyo, the Japanese provided official functions, warm hospitality, an earthquake and so many gifts, including clothes for the continuation of her round-the-world voyage on the *Empress of Japan*, that when she reached America her extra luggage had to be carried on a Stearman escort aircraft.

Her dream of reaching San Francisco for Christmas vanished on a refuelling stop at Medford, when the right wheel of the aircraft collapsed and it made a complete somersault, trapping her again upside down in her straps. It was Christmas Eve, and 'it seemed as though the whole world had collapsed'. Although the damage was repaired by New Year's Eve, Mrs Bruce wished the flight was over: she was 'so tired of flying, flying, flying'. She had still to contend with ice on the wings, a new and alarming experience, as she flew on across the United States in winter in a tiny aircraft with an open cockpit: 'quite an undertaking: in fact, nobody attempts to do it'. In San Francisco, she was given a motor-cycle police escort, sirens blaring, and in San Diego, mistaking a military review for a reception in her honour, she committed the crime of flying through it and nearly knocking off the general's hat. On the way to Chicago, where Al Capone secretly added his signature to the many already scrawled on the Bluebird, the extreme cold affected her reactions: before continuing to drop a flag over her mother's old house, she borrowed fur-lined boots and a flying suit.

With a sense of foreboding – her St Christopher had been torn from the aircraft by a souvenir-hunter – she made a diversion to Washington for an official reception, and had to make a forced landing, chivalrously accompanied by her escort, when she ran out of petrol over the Potomac River. On the way back to New York, she turned the Bluebird over for the third time on landing: with blood pouring from a gash on her forehead, she sat in the mud beside the aircraft and wept. She had at least landed in the best place to have it repaired, right in front of the Glenn Martin Aeroplane Factory. Before she embarked on her second ocean crossing by ship she had a near-miss with a New York skyscraper.

The weather, which she felt had been against her all the way, provided fog in the Channel, but Mrs Bruce felt that she had achieved the status of an international pilot when she was met in the air and escorted from Lympne to Croydon by Amy Johnson and a fleet of Blackburn Bluebirds, and her Bluebird was later exhibited outside Charing Cross underground station.

She had set off with only 40 hours' solo flying behind her, and had taken nearly five months to fly 20,000 miles, crossing no fewer than twenty-three countries in three continents. Two years later, Wiley Post took less than eight days to make a complete round-the-world solo flight in a Lockheed Vega, and Mrs Bruce was one of a team of aviators who proved the possibility of aerial refuelling in Britain, staying in the air for three days and coming down only because of a break in their aircraft's main oil-feed pipe.

Mrs Bruce continued to fly, entering competitions and for a while joining the British Hospitals' Air Pageant flying circus, and even organised her own floodlit flying display at Hampton Court. For several years she operated a company called Air Dispatch, offering a rapid delivery service of 'freight, whether of a fragile nature, ordinary merchandise, machinery parts, or livestock', which soon included an aerial ambulance service. With two de Havilland Rapide twin-engined six-seater aircraft and an Avro 642 which could carry sixteen passengers, the company later gained a contract to carry gold bullion, ran the fastest air service between Paris and London, and was one of the first to carry air hostesses. Mary Bruce also initiated an Inner Circle Airline linking Heston, Hanworth, Gatwick and other airfields: the first passengers were two children who flew daily from Croydon to their school in Hounslow with a £1 2s 6d weekly season ticket.

During the Second World War Air Dispatch had twenty-three aircraft, which were used on army co-operation flying: as this took place at night, Mrs Bruce resumed her early interest in horses, coming second to Colonel Llewellyn at the 1938 Horse of the Year Show and winning the open jumping at Windsor in 1938. She was still adventurous enough many years later when, at the age of seventy-eight, she drove a Ford Ghia Capri at 110 mph round Thruxton race track: 'Going slow always made me feel tired,' she commented.

The Duchess of Bedford

The Duchess of Bedford started flying when she was over sixty; with her private income of £30,000 a year, she could afford to employ a private pilot and a mechanic, own several aircraft, usually more than one at a time, and build a hangar at Woburn. Because of her deafness, which she attributed to an attack of typhoid when she was sixteen and was living with her parents in India, she found listening to other people an almost unbearable strain: in the air, she at last found relief from the buzzing in her ears which accompanied it and which at times was so loud that she described it as being like a steam train rushing through a station.

The daughter of an archdeacon, Mary du Caurroy Tribe married Lord Herbrand Russell, the second son of the ninth Duke of Bedford, in 1888. They became the Duke and Duchess of Bedford five years later, moving to Woburn Abbey and assuming responsibility for an estate of over 3000 acres inside 11 miles of brick wall. Although she fulfilled her social obligations conscientiously, the Duchess preferred to escape on bird-watching expeditions to the solitude of a cottage on the Western Isles or to sea on her luxurious private motor yacht, *Sapphire*. Neither her duties as a duchess, nor as a mother – she had little patience for or interest in her only son – provided a satisfying outlet for her energy. In 1898 she founded a cottage hospital at Woburn, which became a military hospital during the First World War and was her chief interest and occupation for nearly forty years, although during the last ten years of her life flying became almost as important. She summoned her aerial chauffeur at a moment's notice so that she could relax after working in the hospital by taking a flight as others might have taken a walk.

The Duke detested her addiction to flying, but accepted it as he accepted, and indeed supported, her hospital. The Duchess often disappeared for weeks at a time with her pilot on long-distance flights, but when she was at Woburn she made a point of keeping half an hour free every evening for a stroll with her husband, even after a day as full as one recorded in her diary for 1931. By then, she had her pilot's licence and owned both a Gipsy Moth and a Puss Moth; with Flight Lieutenant Allen, her second and favourite pilot, she flew to Brussels and visited two art galleries, returned to Woburn for an afternoon cup of tea, and then they each flew a Moth to Reading, leaving the Puss there for its annual overhaul, before flying back 'in time to walk out with Herbrand, which completed a varied and well-filled day'.

Her first long-distance flight, piloted by Captain Barnard in a Fokker in 1928, started inauspiciously when the aircraft removed some telegraph wires on take-off, and ended with a six-week delay in Persia after the engine was written off by a blocked oil supply, filling the cabin with smoke and covering the windscreen with oil. The Duchess reluctantly returned by ship, but just over a year later, with a Mr Little as navigator, they achieved their ambition of proving 'that a passenger could go to India and back in eight days without undue fatigue' – although few passengers would have enjoyed 'the early starts, the long hours in the air, the pleasant but irregular and rather sparse meals' with quite the same zest as the indomitable Duchess. The passenger accommodation was spartan, and the Duchess, balancing on a cut-down deckchair, spent a quarter of every hour pumping petrol, although she claimed that most of the time she sat comfortably in her 'armchair' and admired the scenery.

Captain Barnard gave her her first flying lesson a month after their return, and a week later she looped the loop. She made her first solo flight in 1930, two days before setting out with the same pilot, navigator and aircraft on another expedition, to the Cape and back. They had breakfast in Lympne and dinner in Oran, in North Africa: 'in these early days of flying, quite an achievement'. The Duchess was allowed to take the controls for as long as one-and-a-half hours at a time, and they reached Cape Town after seven relatively uneventful flying days. On their return journey, they made an emergency landing with a suspected oil leak near Lake Victoria, greeted by a 'weird collection of natives dressed in girdles of beads only' – who found the Duchess' appearance quite as amusing as she did theirs and escorted them through the jungle to a primitive hut apparently reserved for the use of stray aviators.

Their route took them swiftly from a very hot to a very cold climate: it was the cold which caused the greater discomfort, with draughts from the luggage compartment below the cockpit chilling them from the soles of their feet upwards. The Duchess looked on long flights as 'the one perfect rest I get in the year, though at present the non-flying public cannot realise it', but found the heroine's greeting on her return exhausting. Later that year, she was among those to greet Bert Hinkler after he made the first solo crossing of the South Atlantic, and was sure that, like her, he must have had his nerves more stretched by his welcome than by his lone voyage.

In 1931, under the guidance of Flight Lieutenant Allen, who was installed at a salary of £1000 a year in a small house in Woburn Park, the Duchess gained her licence, competed successfully in Brooklands' *concours d'élégance* as pilot-owner, and opened the first All-Women's Flying Meeting at Sywell. In 1933, she and Allen took a month's flying holiday together across Europe and the Middle East. The Duchess, with 50 hours' solo flying behind her, was experienced enough to act as co-pilot, although her position behind the pilot made the role difficult. His bulky figure, in a heavy coat and felt hat, blocked her view both of the landscape and of the instruments, and his weight pressed back on the front seat, behind which she had had a compass fitted. This made any adjustment an awkward operation during which the aircraft would dip or soar alarmingly, and Allen would shout down her speaking tube, always remembering to add to the most irate instruction a polite 'please – Your Grace'.

Flight Lieutenant Allen was killed five months later in an aircraft lent to them for a 'Tour of the Oases' race at Cairo. The Duchess, who had flown more than 1800 hours, most of them with him, felt that she had lost both a skilled pilot and a loyal and trusted friend. Her relationship with her next and last pilot, Flight Lieutenant Preston, was less relaxed, although on their first major expedition, in 1934, they formed the habit of playing three games of backgammon together every evening, sometimes in the most unlikely places. They remained, however, on formal terms: she never used her pilots' Christian names, and they always called her 'Your Grace', even when it seemed as if disaster was inevitable.

There were several such moments on their 1934 flight. Like Lady Heath, they were shot at, unaware of the danger until they landed and discovered two bullet holes in the aircraft's right wing: 'Pretty good marksmanship,' commented the Duchess, 'as we were 3000–4000 feet up and flying at over 100 mph'. Between the Canaries and Marrakesh, they had 'the interesting experience' of facing apparently certain death when the aircraft developed an airlock over a rough sea. The Duchess calmly continued knitting, contemplating death with some disappointment that the flight would be 'wasted' because no one at home would hear about it. It would nevertheless, she thought, be a swift and agreeable way of dying, and she derived private amusement at the contrast between the reality and the inevitable press sensationalism: 'At all events I should have gone down knitting a prosaic sock, and the reporters would have pictured me clinging round Flight Lieutenant Preston's neck and imploring him to save me!'

By the time Preston had worked for her for a year, the Duchess was complaining that her pilot had assumed the attitude of Dictator rather than Adviser. The rift was healed by an article in *Shell News* about a recent flight across the Sahara. 'Shall we go?' Preston asked: two weeks later, they left for a six-week tour which was marred only occasionally by disagreements. The Duchess, with nostalgic memories of her record-breaking flights with Captain Barnard and their dawn take-offs, wanted to be up and away as early as the light would allow; Flight Lieutenant

Preston preferred to leave later after waiting for a weather forecast, thereby, the Duchess insisted, running the risk of poor visibility in the midday heat haze. She was also impatient with his insistence on flying so high that she could not identify birds and animals below her. She nevertheless found it 'an exceedingly interesting flight', in spite of the extreme heat, blowing sand and dust-laden air and the monotonous scenery.

The interest was not confined to the air. At Sokoto, she visited the Emir of Kano in a mud-walled palace incongruously furnished with maple furniture from England. Led by a eunuch through small dark rooms swarming with grubby children, she inspected the harem, where the Emir's six wives lived in almost airless squalor. The heat outside blistered the triplex glass windows of the Puss Moth. It was cooler during an unscheduled overnight stop in open desert after an emergency landing: the Duchess cheerfully helped to fill sacks with stones to anchor the aircraft, then played her usual three games of backgammon before sleeping on the ground.

She celebrated her seventieth birthday later in the year with a short solo flight to London, but in the New Year she cancelled an aerial expedition to Africa because of a 'very bad turn of deafness and noise in the head'. She was feeling increasingly depressed as she envisaged an old age deprived of outlets for her energies: as her eyesight was beginning to fail, she was afraid that she would fail the annual medical test on which her pilot's licence depended. In addition the Duke had told her that he could no longer bear the cost of the hospital.

On 22 March 1937, the Duchess of Bedford took off to bring her solo flying time to 200 hours. More than a week later, after intensive searches, the wing struts of her aircraft were washed up one at a time on the coast, although the route she had said she was taking went nowhere near the sea. There is no way of knowing exactly what happened, but it seems likely that, at the age of seventy-one, the 'Flying Duchess' had calmly chosen the time and the manner of her death.

Although it is tempting to dismiss the flying aristocrats of the 1920s and 1930s as wealthy eccentrics, the publicity which they attracted for aviation certainly helped to promote the acceptance of travel by air. In 1929 Lady Maud Hoare, whose husband Sir

Samuel Hoare was then Secretary of State for Air, flew with him on the Imperial Airways inaugural flight to Karachi and was asked what she felt should be women's contribution to aviation: '. . . to use it, of course, as a quick and easy method of travel,' she replied. Although the speed and ease of air travel were not always apparent in the self-imposed ordeals of Lady Heath or Lady Bailey, or of the adventurous Mrs Victor Bruce and the indomitable 'Flying Duchess', the fact that they were women and that they not only survived, but even enjoyed, long-distance flights in small aircraft did as much as anything else to persuade the general public of the feasibility of air travel.

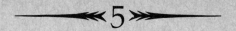

5

American Airmindedness
and Togetherness

The American public, although eager to be entertained and
thrilled by exhibitions of stunt flying, was initially reluctant to
take to the air, in spite of the obvious advantages of overland air
travel across the country's vast distances.

Persuading people that flying was safe became, therefore, a
major preoccupation of aircraft companies and aviation maga-
zines during the 1920s. They used every publicity device to instil
what they called 'airmindedness'. In 1922 a woman was even
flown as an air-mail parcel 2500 miles across America. A short-
lived voluntary Women's Aviation Corps was formed when the
wartime ban on civil aviation was lifted and used to promote
airmindedness at local exhibitions, although this was not the use
intended by its founder, Rodman Wanamaker: the millionaire
Special Police Deputy hoped that it would form part of a permanent aerial police reserve, but the authorities were not yet airminded enough to accept flying traffic controllers and within a
couple of years the WAC was disbanded.

The drive towards airmindedness opened up career possibilities with companies seeking publicity through the appointment
of women demonstrators and promoters. If a mere woman could
fly one of their aircraft, argued the men, it must be particularly
safe. The women were happy to oblige: they too believed in
aviation, and could enjoy free flying in return or afford to pay to
fly by advertising aviation products such as goggles, parachutes
and oil, and writing airminded articles.

Air meetings featuring serious competition flying gradually
gave respectability to aviation as a sport and offset its somewhat
disreputable barnstorming and wing-walking image. Women
were determined to compete, although at first they were barred

from some races. By the time seventy US Department of Commerce pilots' licences had been granted to American women, and forty times as many to men, a survey of male opinion still upheld the belief that women would panic in a crisis, would not be able to make instant decisions, and could not stand having oily faces or clothes.

Even women's flying records were recorded only as 'Miscellaneous Air Performances' until 1929, a year after a woman's endurance flight attempt was first officially monitored. Viola Gentry, protected against the cold of an open cockpit in two flying suits, a woollen bonnet under a fur-lined helmet, gloves and boots, stayed in the air for eight hours in a borrowed Travelair biplane. Her aim had been to stay up for 13 hours, 13 minutes and 13 seconds – thirteen was her lucky number – but she was forced to land when fog and rain turned to ice.

At the beginning of 1929, two young women competed with each other for the record. On 2 January Bobbi Trout, a twenty-three-year-old Hollywood stunt pilot, added more than four hours to Viola Gentry's time in a Golden Eagle, and by flying at night she simultaneously created another women's record. At the end of the same month, Elinor Smith from New York raised it by another hour in an open-cockpit Bird biplane. In February, Bobbi stayed up for 17 hours. Louise Thaden broke into the competition with a flight of just over 22 hours in March, and in April Elinor stayed in the air for nearly 26½ hours in a Bellanca CH monoplane. Although this was considered a surprisingly large aircraft for a girl aged seventeen, she found the assumption that women could handle only light aircraft insulting.

When a Californian businessman offered to sponsor a joint endurance flight with aerial refuelling, Bobbi and Elinor accepted the chance to become the first women to refuel in mid-air and to beat the new men's record of 420 hours (17½ days). When they took off in mid-November, they planned to stay in the air for a month, but had to land almost immediately because the heavy radio with which they were to keep in touch both with the refuelling aircraft and the ground was affecting their Sunbeam's balance. A second attempt without the radio had to be abandoned when part of the rigging gave way. Their third attempt, far from being lucky, ended after 18 hours when Bobbi was soaked in petrol, some of which she swallowed, and had to spend the night in hospital.

At the end of November they at last succeeded in setting a women's record of 42 hours, although this was only one-tenth of the men's record they had hoped to break: they had to land when they ran out of fuel after their petrol supplier, an elderly Curtiss Carrier Pigeon, made a forced landing. They had nevertheless achieved a welcome amount of publicity, actively courted during their preparations: they had posed for photos sitting at the controls, feigning sleep in the makeshift berth on top of the fuel tanks, being massaged beside the aircraft, demonstrating their refuelling technique, and even showing off their outsize radio and microphone.

Bobbi Trout was lucky enough to have a job in aviation as a demonstrator of the Golden Eagle which she flew in her other record attempts. Although she had had her first flight in 1922, when she was sixteen, she had had to wait six years before she could afford to have lessons, saving money while she worked as manageress of her father's filling station. When the manufacturer of the Golden Eagle offered her a job, with flying lessons as part of her salary, she leapt at the opportunity of being paid for what she most wanted to do.

For seventeen-year-old Elinor Smith, publicity was the only means by which she could pay for flying, which had been her sole ambition since the age of six: she claimed to have learned to fly at the age of eight which, although unusually young, was not impossible as there was then no official age limit. Some of her attention-seeking stunts brought more criticism than praise: at the age of fifteen she had her licence suspended for flying under all Manhattan's East River bridges. On one of her many attempts to create new altitude records, she passed out at 25,000 ft when her oxygen tube broke, was unconscious during a four-mile drop and came round only at 2000 ft. It was only lack of money which forced her to abandon the idea of a solo Atlantic flight. She later found a job where she could give publicity to others, as the presenter of a weekly radio programme on aviation, and made her views about discrimination against women well known.

Active discrimination, for example, forced Helen Richey, America's first commercial woman pilot on a scheduled airline, to give up her job. She was appointed by Central Airlines in the early 1930s: the pilots' union at first refused to allow her to fly at all, then restricted her to flying when the weather was considered suitable for a woman. Tired of fighting, and determined not to be merely a fairweather pilot, she resigned, although she continued to fly privately and in competitions.

That women in America were taking up flying and seeking equality in the air was being noted, even if it is difficult to believe that articles such as one which appeared in an aviation magazine in August 1929 were taking their efforts seriously:

> The possibilities of the airplane for the sportsman are unlimited and the smart people have taken to flying with a vengeance. And how the members of the FAIR SEX are jumping at the opportunity to get ahead of their lesser halves! While friend husband is busy at his office, friend wife can be found at the field taking lessons or getting in solo time. For this is one game that can be played by both sexes and it has come into its own right in the midst of the period when equality of sexes is an issue.

The men were nevertheless forced to accept women as serious pilots after their first competitive cross-country event, a Women's Air Derby, organised as the opening attraction of the 1929 National Air Races in Cleveland, Ohio. The race, which carried a $2500 prize, started in Santa Monica, just north of the Mexican border near Los Angeles, and finished over a week later 2800 miles away in front of the main grandstand at Cleveland Municipal Airport. It aroused considerable interest, and was nicknamed the Powder Puff Derby: the women were called Petticoat Pilots, Angels, Sweethearts of the Air, Ladybirds and Flying Flappers. In spite of frivolous comments and even ridicule from the more male chauvinist members of the press, the marathon was recognised as a gruelling test of endurance and flying ability.

Initially, the rules were worded to allow mechanics to accompany the pilots, but they were changed to make the Derby a solo race when a number of Hollywood actresses tried to enter with male pilot-mechanics. To be eligible, the women pilots had to have a licence and at least 100 hours' solo flying time, including 25 hours gained in cross-country flights of 40 miles or more. Of the forty American women who could meet the requirements, eighteen took part, joined by a German, Thea Rasche, and an Australian, Jessie Miller. Thea Rasche was the first German woman to gain a pilot's licence between the wars, when any potentially military flying in Germany was banned. She found flying 'more thrilling than love for a man and far less dangerous' and became Germany's first woman stunt pilot. Jessie Miller had recently completed five months as co-pilot on a 12,500-mile flight from London to Darwin, making her the first woman to fly to Australia: she had set out as a non-flying companion with pilot Bill Lancaster after raising half the money for his aircraft, and learned to fly on the way.

The American entrants included a wealthy socialite, Opal Kunz, who could afford her own aircraft, and at least one woman who flew purely for fun. The majority, among them Amelia Earhart, were involved in aviation professionally, and several were flying aircraft belonging to the companies which employed them: Phoebe Omlie entered in the Monocoupe which she demonstrated; Ruth Nichols worked for the sales department of the Fairchild Airplane and Engine Company and flew a borrowed aircraft; demonstrating the Golden Eagle was part of Bobbi Trout's job. The smallest entrant, who was only 4ft 11in. tall and flew on a pile of pillows, paid for her flying by working both as an estate agent and a film extra. Several were part of husband-and-wife flying teams, earning their living from barnstorming and tuition at flying schools.

The 20 competitors set out at one-minute intervals from Clover Field on 18 August 1929. The first leg was 60 miles to San Bernardino, and for the next eight days the women were in the air by six o'clock in the morning, covering an average of 300 miles a day and preparing for the next day's flight in the evenings, as well as coping with reporters and autograph hunters. Amelia Earhart claimed that hordes of women turned out to gawp at the Powder Puffers and their aircraft, even poking the latter with umbrellas to find out what they were made of.

The week was not without the inevitable problems inherent in such an ambitious undertaking, the danger of which was recognised in the requirement that all entrants must take a gallon of water, a three-day supply of food, and a parachute. The mechanics who were available at each stop managed to complete any necessary repairs overnight and even occasionally during brief day-time stops. There were two unproven suggestions of sabotage, with which Thea Rasche had been anonymously threatened.

None of the women showed any sign of feminine weakness or panic, although Ruth Elder admitted to an unspoken prayer after an unscheduled landing in a field of cattle in her bright red aircraft: 'Please let them be cows.' Blanche Noyes remained admirably calm when she made a forced landing in desert because she could smell smoke, put out a fire in her baggage compartment with sand, and managed to take off and later land again with the loss of a wheel.

In three of the towns on the route, aircraft were a novelty whatever the sex of the pilots. Bulldozers had hastily created airstrips too new to be marked on the road maps which, with unreliable compasses, were the only aid to dead-reckoning navigation. At the start of the race, dustclouds, drifting sand and heat haze made it difficult to identify the landing strips from the air. On the second evening, one of the pilots, who had started late, finished the leg at night: she landed where she saw lights, only to find that she had overshot the Californian border and was in Mexico. The next day, another competitor was following a railway line after losing her way, and also found herself on the wrong side of the border. After intense heat and sandstorms at the beginning of the race, stinging driving rain later made flying both difficult and, for those in open cockpits, uncomfortable. The terrain over which the competitors had to navigate became gradually more demanding, finishing with forested hills and valleys.

When one of the competitors, Marvel Crosson, was killed when she baled out too low for her parachute to open, a newspaper headline claimed that 'Women Have Conclusively Proven That They Cannot Fly'. A Texas oilman by the name of Haliburton stated that 'women have been dependent on men for guidance for so long that when they are put on their resources they are handicapped'. The race manager, Frank Copeland, responded emphatically if inelegantly: 'We wish to thumb our collective nose at Haliburton. There would be no stopping this race.' Fifteen of the women proved him right by reaching Cleveland safely, in spite of the accumulating exhaustion which imposed considerable strain on all the competitors. Three had damaged their planes, but not themselves, and had withdrawn, while one had given up after flying for two days with a temperature which turned out to be a symptom of typhoid. Ruth Nichols, who until then had been in the lead with Amelia Earhart, was forced to retire on the penultimate day when her aircraft left the runway in a strong crosswind, hit a tractor and somersaulted several times before coming to a battered halt.

First into Cleveland was Louise Thaden who, at twenty-three, already held women's endurance, altitude and speed records and had an evangelical belief in aviation and in the significance of the race: 'The successful completion of the Derby was of more import than life or death. . . . We women of the Derby were out to prove that flying was safe.' The cause of airmindedness had been more than adequately served, as well as that of togetherness. The Derby was the first time so many women pilots had been together: for most of them flying was a lonely occupation. Some had never met another pilot of their own sex, as in several states there were no more than one or two women who flew either for work or pleasure, and the nearest woman with whom the interest could be shared might be hundreds of miles away. Amelia Earhart called the race 'the event that started concerted activity among women flyers': by the beginning of November the new togetherness had been consolidated.

In October 1929 four women working as Curtiss demonstration pilots sent a letter to as many other women pilots as they could trace inviting them to a meeting in a hangar at Valley Stream on 2 November. The 26 women who attended the meeting paid a subscription of $1 and decided that the organisation should be open only to women with pilots' licences and should cater for social as well as professional interests. By the next meeting, on 14 December, the total 126 licensed women pilots in the country had been invited to join. All but 27 accepted, and at Amelia Earhart's suggestion the organisation was christened 'the 99s' after the number of founder members, a whimsical title which it has kept ever since whatever the membership. The first president was Amelia Earhart herself, whose fame and energy did much to ensure its success.

Some of the early activities of the 99s were stunts which made a most satisfactory impact on the headlines. On 19 December 1930 four women pilots 'perpetrated an air raid on New York', bombarding the city with leaflets asking for contributions to a women's fund-raising appeal for Salvation Army unemployment relief and advertising a charity supper dance organised by a committee of women flyers. The stunt even had official approval, with an escort of Army, Navy and Police planes, 'making quite an impressive procession in all'. From 1000 ft above New York the voice of Mrs Samuel Clark (not a pilot) was clearly heard as she added an airborne broadcast appeal: 'These wonderful girl flyers have formed a team to help us. Come to the party at the Biltmore this evening and help us go over the top.'

A similar publicity stunt was carried out in September 1933 to advertise the National Reconstruction Act (NRA) designed by President Roosevelt to restore jobs and prosperity after the Depression. Roosevelt had pledged himself at his nomination the previous year to a 'new deal for the American people': the NRA was greeted with a vast parade on the ground and a smaller display in the air. A New York newspaper reported enthusiastically: 'Fifteen feminine flying aces thrilled thousands yesterday with an NRA Air Pageant over Manhattan. Ten planes, led by that of Elinor Smith, carried the flyers through maneuvers and ended with dropping bouquets.' Betty Gillies, one of the other 'feminine flying aces' who took part, was less impressed, and wrote in her diary, 'A very much disorganised formation of pilots happened over NYC [New York City] this morning in honor of the NRA. And I was one of them . . . It was all terrible. Never again!'

The 99s also organised Air Tours, the flying version of caravan rallies. Some were more eventful than others, and one in January 1932 threatened to finish in disarray. Several planes took off from Hicksville, on Long Island, to fly in stages to Miami. On their second day, as they were approaching Pinehurst in North Carolina, there was a sudden deterioration in the weather. Four of the aircraft reached Pinehurst Airport without mishap. The others landed in some extraordinary places: one on top of trees outside a sanatorium; another, a Monocoupe, on rough pasture from which several tree stumps had to be removed before it could be retrieved. Two landed on a golf course: one on a fairway and the other, the accompanying 'Official Airplane' with all the pilots' luggage and the editor of the *New York Times* on board, on a driving range. Somehow they were all able to fly on to meet the rest at Pinehurst Airport the next day.

Among the less spectacular but more practical achievements of the 99s was the introduction, at the suggestion of Phoebe Fairgrave Omlie, of aerial markings. She felt that 'an air route without markings was like a highway without signs', and was in a good position to do something about the lack as, from 1934 until 1936, she was on the National Advisory Committee for Aeronautics. She gained powerful support: it was Eleanor Roosevelt's idea that the Bureau of Air Commerce should hire women pilots to find suitable buildings and negotiate with local officials and owners for permission to paint on their roofs. In orange letters large enough to be legible at 3000 ft, place names as well as distances and directions to airfields soon started to appear approximately 15 miles apart on every air route in the country. Altogether 16,000 huge signs were painted on rooftops to guide pilots over the vast open spaces of America. With the limited navigational equipment then available, it was only surprising that no one had thought of it before.

Phoebe Fairgrave Omlie relinquished her position on the National Advisory Committee for Aeronautics to support Roosevelt's re-election campaign with a 10,000-mile tour. She had come a long way in sixteen years since her wing-walking and barnstorming days, and in 1927 was the first woman to be granted a transport licence. During a flood disaster in the Mississippi Valley she was celebrated as a local heroine, flying medical supplies and food into the stricken area, and in 1928 was the first woman to complete a National Air Reliability Tour in which she flew over 5000 miles and visited thirteen states in a month. In 1931, when for the first time women competed against men in the National Air Races in Cleveland, she won $2500 and a car for the highest points overall.

It was the US press which decided who would become a household name and who would not. The sudden rise to fame of Ruth Nichols followed much the same pattern as that of Amelia Earhart, who became famous as the first woman to be flown safely across the Atlantic. Ruth Nichols, who had started flying in 1922, was unknown until 1928, when her flying instructor, Harry Rogers, invited her to be his co-pilot on a pioneering flight from New York to Miami. They created a twelve-hour record, almost entirely, as Ruth insisted, due to Rogers. By this time, when a man broke or made any but the most exotic record it had little news value, and it was Ruth who found herself the centre of attention with headlines like 'Flying Deb Pioneers New York-

Miami Hop'. To call her a 'flying deb' was not entirely fair, although her background provided some excuse. She was the daughter of a wealthy New York stockbroker and had been educated at an exclusive and expensive private establishment before going to medical school. Her parents were equally opposed to their daughter studying medicine or taking to the air, and wanted her to launch herself into New York society as befitted a young lady of her status. A compromise was reached with a job in a bank.

Ruth's sudden flying success at twenty-seven enabled her to embark on a career in aviation as a saleswoman for the Fairchild Company. At the same time, she started planning and preparing herself to conquer the Atlantic with a solo flight to Paris. Her outfit of a matching purple leather flying suit and purple helmet helped to maintain the glamorous image of a flying deb, although her record flights during the next two years were made in a borrowed aircraft, a Lockheed Vega with the Indian name *Akita* (Discovery).

She started her preparation by entering the 1929 Air Derby and continued it by flying across America and back, breaking both the east-west and west-east women's speed records, in 1930. In March the next year, after having had the Vega stripped down, she put longjohns and four sweaters under her flying suit and broke the women's altitude record, reaching nearly 20,000ft above Jersey City. She climbed until she had only a five-gallon reserve tank of fuel left, when her engine stopped and she 'dropped 5000 ft like a stone', landing with her tanks nearly empty and relieved not to have ruptured her eardrums. Her makeshift oxygen supply caused her some alarm as well as considerable pain: she simply sucked the oxygen from a tube inserted into the tank, and while she was at over 20,000 ft this meant that she was breathing it in at well below freezing point. Later the same year, she beat Amelia Earhart's 3-km speed record by nearly 30 mph.

On 2 June 1931, with extra fuel tanks and navigational equipment, Ruth at last set out for Paris, but crashed on landing to refuel at St John, New Brunswick. Instead of flying the Atlantic she found herself in hospital with five broken vertebrae and a doctor's ban on flying for at least a year. Ignoring medical instructions, she arranged for the aircraft to be repaired and only a few weeks after the accident drove a car for thirteen hours to prove that she was fit to make another transatlantic attempt. This time it was the weather which stopped her, and she reluctantly postponed the project. Instead, since she had been ready for a long tough flight, she decided to try for the women's endurance record by flying non-stop from California to New York. Out of plaster at last, she was however still encased in an unyieldingly restricting steel corset. To allow her to take the weight off her thighs and lower spine while she was flying, she had loops attached to the cabin roof. It was, however, because of strong crosswinds rather than the discomfort caused by her metal straitjacket that she landed short of her original destination, at Louisville, setting a new record of just under 2000 miles non-stop, 100 miles more than any woman had managed before.

As she was about to take off the next day, the exhaust set fire to a leaking fuel valve. If she had not landed at Louisville, it is likely that this would have happened in the air, and that she would not have been able to escape before the petrol tank exploded: as it was, she managed to stop the aircraft and scramble to safety just in time, but *Akita* was written off. By this time Ruth's friend and rival Amelia Earhart had beaten her across the Atlantic: the determination of both women encouraged and inspired many younger pilots. Ruth was the first woman director of a large aviation corporation, and flew all over America to publicise an organisation called Aviation Country Clubs: her solo promotional tour took her to 98 cities, covered 12,000 miles and made her the first woman to land in all of America's then 48 states. She was also one of half a dozen women pilots, including Amelia Earhart, who wrote regular articles about aviation.

Ruth Elder, who came fifth in the 1929 Derby, was another of the early 99ers whose name was already widely known. In her case the publicity was deliberately sought but not given for quite the reasons intended. She was still a student pilot when she announced, in August 1927, that she was going to fly the Atlantic and do it, moreover, in the same year as Lindbergh's much-publicised solo crossing. After persuading a group of business-men to sponsor her and her instructor, George Haldeman, to accompany her as co-pilot, she ignored all advice to wait for a better chance of good weather in the spring. The flight was partly a publicity stunt for a Stinson Detroiter aircraft christened

American Girl, but Ruth Elder gave two more personal reasons for wishing to make it: she wished to be the first woman to fly the Atlantic, and she wanted to buy an evening dress in Paris.

American Girl left America for Europe on 11 October, taking a southern route. After flying for more than 2500 miles, a record distance over water, George Haldenman and Ruth were forced to put the aircraft down near the Azores when the oil pressure dropped. Their choice of route saved them, as it took them over shipping lanes: they were able to ditch near a Dutch freighter, which picked them up before they had time to test the efficiency of their bulky new rubber safety suits. Minutes later, *American Girl* exploded. Ruth had done none of the flying herself as the weather had been too bad for someone so inexperienced, but she had earned the unwished-for distinction of being the first woman pilot to be rescued by ship from the Atlantic. Had the flight succeeded, the praise reserved for Amelia Earhart only a few months later would have been Ruth Elder's. As it failed, it was criticised. A woman sociologist, Katherine Davis, allied herself with male attitudes when she stated that 'There is no woman alive today . . . equipped for such a flight.'

Not all 99ers were as newsworthy as Amelia Earhart and the two Ruths, but during the 1930s it became quite usual for women to compete with men, and win: there were few races in which American women pilots competed without being placed. In 1936, Louise Thaden followed her 1929 Derby success by being the first woman to win the Bendix race from Los Angeles to Cleveland, with ex-actress Blanche Noyes as co-pilot. The Bendix Trophy brought with it $7000, and recognition both for the two women and for the Beechcraft Staggerwing with which they had beaten the men's powerful racing and twin-engined aircraft: Louise Thaden had always claimed that women were 'innately better pilots than men'. She had been given the chance to fly in 1926 by Walter Beech, at whose Travel Air factory she had spent so much time while she was working for a Kansas coal company that he had offered her a job, with flying lessons as part of her salary. Beech was repaid by the publicity which she gained for his aircraft with her frequent record-breaking and competition-winning flights.

The need to campaign for airmindedness became gradually less important as air travel gained momentum. The fight for women's rights in aviation eventually started to produce results, and by 1979, fifty years after the start of the concerted effort among women pilots, there were an estimated 20,000 women pilots in America. These included commercial pilots, women naval officer pilots, women with helicopter ratings in the army and two Air-Sea Rescue helicopter pilots. The 99s had increased to around 6000 members and spread to nine other countries. While women were proving their ability to fly well and safely, furthering commercial interests in aviation by inspiring air-mindedness and writing articles with the same purpose, many of the 99s' activities on the ground during their first half-century make the organisation sound a little like the British Women's Institute with wings. They talked and demonstrated at schools, fought against airport closures, and took 'their knowledge and love of aviation to the community at large'. They also fried sausages for pancake breakfasts, made cookies for airport controllers, 'joined the bunch for lunch at your favourite fly-in restaurant', filled 'goodie bags' and washed aircraft to raise money.

When 600 women, including several of the founder members, celebrated the 99s' Golden Jubilee at Albany, New York, in 1979, the achievements of women pilots were eulogised by Senator Barry Goldwater. He hailed 'the great accomplishment of women pilots throughout the last half century' and 'the important and lasting contribution' which they had made to 'the development of aerospace pursuits'. Amelia Earhart had put it more succinctly fifty years earlier: 'If enough of us keep trying, we'll get someplace.'

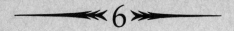

Amelia Earhart:
America's Winged Legend

When Amelia Earhart was the first woman passenger on a transatlantic flight, in June 1928, she described herself as being as useful as a sack of potatoes, but was treated as a heroine. Her subsequent career in aviation, which lasted less than ten years, created a legend to which her death merely added an element of mystery.

By the time she started flying, at the age of twenty-three, Amelia had an independent and feminist outlook, and an aversion to domesticity and alcohol. Her father, a lawyer, had become an alcoholic, and the family's life had for years been unsettled and lonely. She was also a confirmed pacifist: when she was twenty she had thrown herself into nursing at the Spadina Military Hospital in Toronto, where she was constantly faced with the suffering inflicted by war. She left Toronto with a sense of 'the inevitability of flying' as 'one of the few worthwhile things that emerged from the misery of war': she had often watched while young pilots did training manoeuvres at the nearby military airfield, although regulations prevented any of her pilot friends from taking her into the air.

Her nursing experiences exhausted her emotionally and physically, and afterwards she succumbed to severe sinusitis which was to recur periodically at times of stress. After her convalescence, she enrolled as a medical student at Columbia University, but soon decided that she did not want to become a doctor. She opted instead for medical research, and joined her parents in Los Angeles, where, at an airshow at Long Beach in 1920, she persuaded her father to pay $10 for her first flight. She felt no greater emotion than interest, and surprise that the engine was so noisy, that there was no sensation of speed and that the ground looked so unfamiliar from 2000 ft. Nevertheless, Amelia knew by the time she landed that somehow she had to find the money for flying lessons: at least $500 for a dozen lessons.

She chose to start her lessons with a woman instructor, Neta Snook. After only two-and-a-half hours' tuition she bought a $2000 Kinner sportsplane, helped by her mother who did not share the view of the rest of her relatives that flying was 'decidedly queer'. Her pilot friends advised her against the Kinner's small and experimental air-cooled engine, but Amelia found it easier to handle and more fun than larger aircraft: it had the added advantage of free hangar space and mechanical assistance in return for demonstration flights. For the next few years, she took a series of jobs to pay for her flying, and spent her free time at the airfield at Long Beach Boulevard, dressed like the male pilots in long boots, breeches and a leather jacket to which she had deliberately given a suitably used look.

Flying was an escape from the continuing tension at home, and a way of showing her independence from Sam Chapman, her suitor and a lodger at her parents' house. Sam wanted a conventional marriage, but Amelia made it clear that she would not succumb to the restrictions this would impose on her freedom with anyone, although she enjoyed his company and occasionally went with him to left-wing meetings which strengthened her feminist tendencies.

In 1922 Amelia made her first solo flight, and within a few weeks created a new women's altitude record of 14,000 ft in her bright yellow biplane. A few weeks later, this was bettered by Ruth Nichols. Amelia's immediate attempt to regain the lead was her first and almost her last experience of blind flying in extreme weather conditions, without instruments. Protected only by goggles in the open cockpit, she found herself flying at 11,000 ft in sleet and at 12,000 ft in dense fog, disorientated by her inability to see or to sense the aircraft's position. After spinning down through the murk, she broke cloud at 3000 ft and landed safely: as a more experienced pilot commented, she was lucky that the fog bank had not stretched to ground-level.

In 1924 Amelia's parents were divorced. She sold her aircraft to a young man who, watched by her and W.G. Kinner, promptly wrote it off and killed himself by attempting to demonstrate his skill in low aerobatics. Her own flying came to a temporary halt with another severe attack of sinus trouble, possibly brought on by the rigours of flying in an open cockpit and exacerbated by stress after the four-year disintegration of her parents' marriage. The Kinner was replaced by an equally brightly coloured touring car, christened the 'Yellow Peril', in which Amelia and her mother made a leisurely 7000-mile move to Boston, where she went into hospital for an operation to alleviate the almost constant headaches and nasal pain from which she had been suffering.

With no aircraft to pay for, she resumed her medical studies at Columbia but again dropped out. She next tried teaching, then in 1926 became an unqualified but enthusiastic residential social worker at Denison House, an immigrant settlement centre in Boston. She seemed at last to have found her vocation, but at the same time was keeping up her interest in flying on her income of $60 a month. By September 1927, when she wrote to her fellow pilot Ruth Nichols suggesting an association of women pilots and calling herself 'a social worker who flies for sport and . . . the director of an aeronautical concern' (she had a small investment in an airfield), she had accumulated 500 flying hours.

Amelia's life was changed by an unexpected telephone call in April 1928 inviting her to join a transatlantic flight as a passenger. Always ready to accept a challenge, especially if it involved flying, she was interviewed in New York. She was to join the crew of the *Friendship*, pilot Wilmer (Bill) Stultz and navigator Louis Gordon, as the first woman to fly the Atlantic, although she would have no active role. Stultz was to be paid $20,000 and Gordon $5000: Amelia would go 'for the fun of it'.

The Atlantic was still one of the great obstacles blocking international air services, although it had been flown six times since its defeat by Alcock and Whitten-Brown in 1919. Charles Lindbergh had made the first solo crossing in the *Spirit of St Louis* in 1927, a year in which nineteen people died in unsuccessful attempts and three women failed to become the 'First Woman to Fly the Atlantic', all as passengers with male crews. The aircraft of two, British-born Princess Anne Loewenstein-Wertheim and American Frances Grayson, were each seen once after take-off then vanished. Bill Stultz was originally to have been Mrs Grayson's pilot, but had withdrawn because he lacked confidence in her aircraft. The third, Ruth Elder, had been rescued by a passing ship after coming down in the Atlantic. The month before Amelia was asked to join the *Friendship*, the Hon. Elsie Mackay and a hired pilot disappeared while they were trying to fly the Atlantic in the more difficult direction, from east to west.

Considerable advance publicity had been given to all these abortive expeditions, but Amelia confided only in one colleague and in Sam Chapman, whose proposals of marriage she still persisted in refusing. This secrecy was in keeping with the wishes of the flight's organisers, who wanted to avoid the hassle and distraction which would follow any public announcement. The flight's self-appointed publicity officer, publisher George Putnam, nevertheless intended to make as much capital as possible from its success: he had published *We*, Charles Lindbergh's hastily written account of his solo achievement, and hoped to secure another bestseller with a companion book by the first woman across the Atlantic.

The expedition was financed by Mrs Guest, a wealthy American whose English husband, Captain F.E. Guest, had been Britain's Secretary of State for Air in 1921–2. She had bought the three-engined Fokker from Commander Richard Byrd, who had already led an aerial expedition over the North Pole and had planned to use it for a South Pole flight. When Mrs Guest's family and friends dissuaded her from making the flight herself, George Putnam's first task was to find a suitably presentable young American woman pilot to take her place. Amelia Earhart, 'a young social worker who flies', was suggested by a retired Rear Admiral from Boston.

Amelia saw the *Friendship* only twice during the preparations, but was immediately impressed by its golden wings with their 72-ft span. The fuselage was painted bright orange, so that if a search should be necessary it would be clearly distinguishable from its surroundings, and the wheels had been replaced by experimental pontoons, making it the first trimotored seaplane.

As soon as the *Friendship* had taken off from Boston on 3 June, George Putnam called a news conference: the *Christian Science Monitor* carried a story under the headline 'Boston woman flies into dawn on surprise Atlantic trip'. This infuriated an American heiress called Mabel Boll, who had set her heart on being the first woman across the Atlantic, with Bill Stultz as her pilot. For a year, Mabel Boll had been trying to buy a pilot, following first one and then another between New York, Paris and London with her offer of $25,000 and eventually even flying to Cuba with Stultz. No one wanted to pilot 'the Diamond Queen' on a transatlantic flight.

The announcement of the departure proved premature. Fog forced the crew of the *Friendship* down for an overnight stop in Halifax before they could continue to their starting point, Trepassey Bay, where they had a two-week wait for fair weather. Amelia had only the clothes she was wearing, and had to borrow a flannel nightdress: her luggage was a small army knapsack holding her toothbrush, fresh handkerchiefs, cold cream and a comb, although her short tousled hair never looked as if she combed it. A Boston newspaper implied that she was afraid of the long and dangerous flight ahead, but to leave without a good forecast for the next 48 hours could have been suicidal, and to take off from rough water was impossible.

Mabel Boll was delighted at the delay. She had at last found a pilot willing to take her and her $25,000 and still hoped to be first across, even suggesting an unofficial race. Far from wanting to enter a flying contest, Amelia was in the only too familiar position of having to depend on a compulsive drinker. Bill Stultz was drinking more and more heavily with the growing tension, frustration and boredom: when the favourable weather report at last came through, he had to be hauled drunk into the aircraft and Amelia removed a bottle of brandy she found hidden in the cabin.

The *Friendship* took off on 17 June 1928 from the water of Trepassey Bay and flew for 40 hours 20 minutes. Amelia, wearing her usual and by now genuinely well-worn flying clothes, kept the logbook, with a combination of technical detail and personal comment and description, but was very conscious of being merely a passenger. It was an uncomfortable and cold journey, as she crouched at the navigation table behind the extra fuel tanks, snatching a few minutes sleep in the same awkward position and occasionally going forward into the cockpit so that she could enjoy the view, although there was little enough to see most of the time except drifting fog. The three 225-hp Wright Whirlwind engines throbbed comfortingly above the frequently invisible Atlantic, although there were some anxious moments when they were flying under a 300-ft cloud ceiling.

At dawn, with the radio dead, they bombarded a ship with

messages weighted with oranges, hoping that their exact position would be painted on deck. They were exhausted after 19 hours in the air, and Stultz estimated that they had an hour's fuel left: they had had to jettison spare cans to lighten the aircraft on take-off. Half an hour later, they flew over fishing boats and landed in a sheltered bay: they had missed Ireland completely in the murk and were in Burry Port in Wales.

After an hour's early morning peace before anyone took any notice of the seaplane in the bay, the reception from the British press and public was overwhelming, first in Wales and then, after a night's badly needed sleep and the first bath since leaving Boston, in Southampton. Mrs Guest and the Lady Mayor, among many other important and less elevated persons, provided a rapturous reception as the *Friendship* touched down on Southampton Water and the crew was taken ashore by launch. Tugboats whistled, foghorns blared, the crowds cheered, and in spite of Amelia's repeated insistence that the success of the flight had had nothing to do with her, she was the one everyone wanted to see, to touch, to talk to.

Her first visit to England was a two-week 'jumble of teas, theatres, speech-making, exhibition tennis, polo and Parliament'. It included dancing with the Prince of Wales, and shopping at Selfridge's, where she was not allowed to pay. She talked about social work with Lady Astor, visited Toynbee Hall, on which Denison House was modelled, and bought the Avro Avian monoplane in which Lady Heath had flown solo from Cape Town to London.

There was some criticism to dilute the praise. In an article in the *Church Times*, headlined 'Read, Mark, Learn', Amelia was sarcastically congratulated on escaping the fate of the three women who had lost their lives in attempting the same feat. She was, the article continued, an international heroine 'simply and solely because, owing to good luck and an airman's skill and efficiency, she is the first woman to travel from America to Europe by air': precisely the point that Amelia herself was trying to make, although she did not agree that the pilot's anxiety 'must have been vastly increased by the fact that he was carrying a woman passenger', nor did she appreciate being told that 'her presence added no more to the achievement than if the passenger had been a sheep'.

In America, Amelia found she was public property, manipulated by George Putnam, who had organised an exhausting week of receptions in New York, Boston and Chicago. It was at the Putnam family home in Rye that Amelia was at last allowed a few weeks of privacy, as well as considerable space and elegance, to write her account of the flight. She found Putnam 'a fascinating man', and dedicated her story of the *Friendship* flight to his wife Dorothy, mother of his two sons. George Putnam decided which invitations 'Lady Lindy' should accept – her physical resemblance to Charles Lindbergh was remarkable – and organised lecture schedules with up to twenty-seven engagements a month.

The instant heroine became an established and respected spokeswoman for aviation, deeply committed to the potential of air travel and to persuading as many people as possible that flying was not only fun but also safe, and that its future could include a role for women as pilots, in ancillary positions and as passengers. She saw her new cause as a form of social work. At the same time, she had found a career which satisfied her need for independence and constant change, and a way of financing her flying: her various activities, under George Putnam's guidance, earned $50,000 in the first few months.

As an associate editor for *Cosmopolitan*, Amelia wrote articles on such themes as 'Shall You Let Your Daughter Fly?', 'Is It Safe For You To Fly?' and 'What Miss Earhart Thinks When She's Flying'. *McCall Magazine*'s offer of a post as aviation editor was withdrawn when a Lucky Strike cigarette advertisement carrying her name and picture resulted in some not unreasonable criticism. A non-smoker, Amelia squared her conscience by donating her fee of $1500 to Commander Byrd for his Antarctic expedition. Her work for Transcontinental Air Transport was less controversial: she was to make promotional flights and generally stimulate interest in scheduled passenger services.

In September 1928 Amelia flew in Lady Heath's old Avian to Los Angeles to visit her father and go to the National Air Races, and then back to New York. The trip, made in several stages, was another First: no other woman had made a solo return flight across the continent. The lack of navigational aids made distinguishing one small town from another at 100 mph a problem, and she felt that crossing Texas was almost like flying the Atlantic. She created a sensation by landing on the main street of a small town, Hobbs, fortunately free of traffic, after her map was blown out of the cockpit. A few years later she campaigned for aerial markings. In August 1929 she competed in the first Women's Air Derby in a new Lockheed Vega, an aircraft which had fulfilled its test pilot's prediction two years earlier that it would sell 'like hotcakes', and by November she had become involved with the organisation of the 99s, of which she was elected founder president. In the same month, she broke the women's speed record.

Although she was still determined not to limit her freedom by marriage, Amelia eventually gave in to the persistence of George Putnam, who was divorced in 1930 and claimed to have proposed to her six times between 1928 and late 1930. Her friends found him too aggressive and her mother disapproved both because he was divorced and also because he was twelve years older than Amelia. The wedding, in February 1931, was a quiet and private affair at George's mother's house. Amelia kept it deliberately low key by wearing an old brown suit and no hat.

On their wedding day Amelia handed her bridegroom a letter in which she had set out her marriage terms, telling him that she felt 'the move just now to be as foolish as anything I could do'. She absolved both from 'any medieval code of faithfulness', and asked that neither should 'interfere with the other's work or play, nor let the world see private joys or disagreements'. Her conditions even included an option to cancel the partnership after a year. As much a business arrangement as a marriage, it suited both AE and GP, as they called each other. A fortnight after the wedding Amelia wrote to her mother: 'I am much happier than I ever expected I could be in that state. I believe the whole thing was for the best. Of course I go on as before as far as business is concerned.'

Business as usual included flying a Beech-Nut autogiro to promote both the machine and the chewing gum of that name, at the same time creating and breaking more records. GP had sold his interest in the family publishing firm so that he could devote more time to the promotion of his wife's career: reporters complained that it was often impossible to obtain a few words from Amelia Earhart without a few hundred from George Putnam, whose entrepreneurial instincts resulted in airline luggage, women's sportswear, casual and formal suits, stationery, and even an automobile engine all named after her.

Until she had piloted herself across the Atlantic, Amelia felt that her reputation was built on false pretences. In 1932, she and GP agreed that she was ready. She started two months of preparations with the encouragement and technical expertise of Bernt Balchen, a Norwegian pioneering pilot who had flown with Amundsen's 1926 North Pole flight and as pilot on Byrd's 1929 South Pole expedition. Her three-year-old Vega was adapted to give it a cruising range of 3000 miles, with new ailerons, a new engine and extra fuel tanks as well as two new compasses, a drift indicator and a directional girocompass. Amelia took off from Harbor Grace in Newfoundland on the evening of Friday 20 May 1932, watched by Balchen: 'She looks at me with a small lonely smile and says, "Do you think I can make it?" and I grin back: "You bet." She crawls calmly into the cockpit of the big empty airplane, starts the engine, runs it up, checks the mags, and nods her head. We pull the chocks, and she is off.'

For the first few hours she enjoyed the beauty and solitude as she flew through a lingering sunset and into moonlight over a bank of clouds at about 12,000 ft. Then things began to go wrong. The altimeter failed. A violent electrical storm buffeted the aircraft. The wings started to ice up, and she came out of the resulting spin only when the barograph had recorded 'an almost vertical drop of 3000 ft'. Flames from a crack in the engine manifold welding looked alarming in the dark: by dawn it was vibrating badly and Amelia decided to land in Ireland instead of continuing to Paris. Exactly five years after Lindbergh's solo Atlantic crossing, Amelia Earhart startled the cows as she landed in a gently sloping pasture a few miles outside Londonderry. She had been flying for over fifteen hours, during which her only nourishment was some tomato juice sipped through a straw, and she had created three records: first transatlantic flight piloted by a woman, first solo crossing for a woman, and the fastest crossing by anyone.

It seemed very quiet and safe as she sat, exhausted, in the cockpit of the Vega wondering where she was. Even in such an out-of-the-way place, a crowd soon gathered. As usual, the scale and warmth of her reception took Amelia by surprise, and within twenty-four hours she gratified GP by summoning his assistance.

Until he arrived, Paramount News took over, flying her to London, where she stayed with the American Ambassador and visited Lady Astor, who offered to lend her a nightdress, at Cliveden. As usual, she had travelled with only the clothes she flew in, so she treated herself to a complete new outfit from Selfridge's, where she signed a plate glass window with a diamond-tipped pen and where her aircraft was displayed on the ground floor.

First in London, and then, with GP, in Paris, Rome and Brussels, Amelia was showered with honours. She met the Pope, Mussolini, and the King and Queen of Belgium. Three aircraft dropped flowers on the liner *Ile de France* as she left Europe from Le Havre. In America, she was guest of honour at a formal dinner at the White House, followed by a presentation of the National Geographic Society's special gold medal, never before awarded to a woman. The proceedings were broadcast coast to coast over NBC's thirty-eight radio stations. In her speech Amelia made light of her achievements, but hoped that these might have 'meant something to women in aviation'. President Hoover's wife expressed the feelings of thousands when she said how 'nice' it was that the woman to fly the Atlantic alone and so 'represent America before the world' should be 'such a person as Miss Earhart . . . poised, well-bred, lovely to look at, and so intelligent and sincere.'

AE's reputation and GP's inclination brought them instant friends in high places, from the Roosevelts down. Eleanor Roosevelt and Amelia had much in common: both were ex-social workers, both tall, both lived in the constant glare of publicity, both were adventurous and independent. Mrs Roosevelt even hoped that Amelia would teach her to fly, inspired by a nocturnal flight together in evening dress, until her husband put his presidential foot down. Most of Amelia's other friends were fellow pilots: Anne and Charles Lindbergh, many of the members of the 99s, Amy Johnson, who recuperated after a crash at the Putnam's Rye house in 1933, and Jacqueline Cochran, with whom she also shared an interest in extrasensory perception. When the Putnams moved to Hollywood in 1934, they added some of the top people in show business to their circle, and Amelia and Mary Pickford seriously considered a joint film project.

Another Hollywood neighbour was Paul Mantz, an ex-army pilot with a fleet of planes used in films and a flying Honeymoon Express to Yuma and Las Vegas for eloping couples. His wife Myrtle later cited Amelia in divorce proceedings, probably without justification, although as Paul became Amelia's adviser for her long-distance flights she spent more time in his company than with GP. When questioned about her marriage, all she would say was that 'our work and our play are much together'.

She had conquered the Atlantic for a non-stop flight across the continent. Now she wanted to try the Pacific, with a flight between Hawaii and the mainland, a route previously fraught with disaster. Both the army and the navy attempted to prevent the flight, and the mainland press accused Amelia of using her influence to further the sugar interests of a group of Hawaii businessmen who had promised $10,000 sponsorship. When the sponsors threatened to withdraw their support, Amelia accused them of cowardice, and told them she would make the flight with or without them. They honoured their promise.

It was the longest non-stop flight she had ever undertaken, but once in the air she had no problems. Her exhaustion when she landed at Oakland took her by surprise, but she immediately started planning another major record-breaking project, a goodwill flight to Mexico and a return to New York by the shortest route, across the Gulf of Mexico. George Putnam and the Mexicans came up with the idea of financing the adventure through the sale of commemorative stamps. The overprinting of the words *Amelia Earhart, Vuelo de buena voluntad Mexico 1935* on 780 stamps was supervised by GP, who had to make a public statement to satisfy querulous stamp dealers of the authenticity of the stamps and the legality of raising funds in this manner.

By the time the Mexican return flight had been successfully completed, Amelia had made three strenuous long-distance flights in one year, with all the work of planning and attention to detail each involved, as well as her usual non-stop lecture tour. It was hardly surprising that her sinuses should start 'kicking up' again, followed by a pain which she put down at first to a strained muscle but eventually had to admit was pleurisy. Instead of resting, she was soon working with Paul Mantz on various joint ventures, lecturing, and fulfilling new commitments as Purdue

University's occasional resident aviatrix and careers' adviser for the female students.

Amelia's next self-imposed challenge was a round-the-world flight at the Equator, never before attempted. She had decided not to risk any more long sea crossings with a single engine, but Purdue University put $50,000 into an 'Amelia Earhart Fund' for a 'flying laboratory': a new Lockheed Electra was ordered, the most advanced twin-engine dual-controlled civil aircraft available and much in demand for passenger services with its seating for ten in addition to two pilots.

Under Paul Mantz's supervision, the passenger seats were replaced by every imaginable useful device, finishing with more than a hundred dials and levers on the cockpit control panel. The top speed of the production model was 210 mph with a ceiling of 27,000 ft. Her test flight of the new 'laboratory' was given full front-page coverage on 22 July 1936 in the *Los Angeles Times*. Blind flying instruments, a Sperry autogiro robot pilot, a fuel minimiser, wind de-icers, radio homing and two-way sending devices had, the paper reported, all been fitted to the aircraft which now needed only the addition of a dozen extra fuel tanks, three on each wing and six in the fuselage, to be ready for 'a possible non-stop flight of some 4500 miles'.

The Electra was officially presented on Amelia's thirty-ninth birthday. In her speech of thanks she said that she intended to use it to 'produce practical results for the future of commercial flying and for the women who may want to fly tomorrow's planes'. The application to the Los Angeles Department of Commerce for an aircraft licence stated its use as 'long-distance flights and research'. From the end of July until March the following year, Amelia's programme was busier than usual even without the many hours she spent practising on Paul Mantz's Link blind flying trainer and familiarising herself with the complexity of the Lockheed's instruments.

Her route was to be from Oakland, California, to Honolulu, then to Howland Island, a speck in the middle of the Pacific, and on to Australia. From Australia she would fly across India to Arabia, then across Africa and the South Atlantic to Brazil, eventually doing the last leg, to New York, alone. GP took charge of obtaining visas for every country to be visited and permission to fly over others, as well as arranging for spare parts and fuel at every stopping point, and raised $25,000 through the sale of 10,000 letter covers which Amelia would carry with her and mail to collectors en route.

She was to take a navigator, Captain Manning, who had given her some navigational instruction during the sea return from the *Friendship* flight. Amelia and George spent the New Year with Floyd and Jacqueline Odlum, a rare occasion to relax privately, swim, ride and sunbathe at the Odlum ranch: Jacqueline was worried about the long uncharted distances over water, and feared that Captain Manning's excellent maritime navigational ability was not up to aerial high-speed celestial navigation. Amelia, convinced of her friend's extrasensory abilities, tested him by circling several times some distance offshore. On the return course to Los Angeles he was 200 miles out, more than enough to mean the difference between life and death. Captain Frederick Noonan agreed to join them at least for the most hazardous leg, that across the Pacific to Howland Island. An ex-sailor, he had been flying since the end of the First World War, had been chief inspector of all Pan-American's airports and had had the task of charting their new Far East Pacific routes, but had been relieved of his duties because he drank too much.

In February 1937, contrary to her usual custom of complete secrecy until the last minute, Amelia announced her world trip at a New York press conference arranged by George Putnam, who used every excuse for further publicity. More than once, a test flight was rumoured to be the big take-off, but when she eventually left, in the early hours of 17 March, everyone was asleep except those immediately involved and one press photographer who photographed the Electra from a rented aircraft as she passed over the Golden Bridge Gate. The flight to Honolulu set a record for the east to west crossing of just under 16 hours, but the next leg aborted on take-off when the heavily loaded Electra crashed on the runway. Amelia herself was not sure what happened, but arranged for the aircraft to be shipped home for repairs and told reporters that she would be back for another try.

In spite of the estimate of at least $25,000 and five weeks' work to put the Electra back in the air, Amelia never seriously thought of abandoning the flight. Another $25,000 would be needed for all the incidental costs of rearranging flight authorisations and fuel and spare supplies, and she felt that she was mortgaging her future. 'But what', she asked, 'are futures for?' Her immediate plans were assisted financially by a number of generous contributions: the Odlums gave her 'a substantial sum', Admiral Byrd sent $1500, a repayment of the proceeds from her notorious cigarette advertisement, the Lockheed mechanics refused payment for a Sunday devoted to finishing the repairs to the aircraft on time, and Bernard Baruch, a self-made financier and Presidential adviser, sent $2500 just because he admired Amelia's 'everlasting guts'.

Because of the prevailing weather conditions later in the year, Amelia reversed her original route. Fred Noonan was to be her navigator, staying with her the whole way, although George wanted her to fly at least the last leg solo for maximum publicity. She was, however, more concerned about safety than publicity, and this time insisted that there should be no public announcement about the departure until after take-off.

The most difficult part of the flight would be finding Howland, a flat island only two miles by half a mile. Amelia's decision to dispense with a trailing radio antenna meant that she would be out of radio range for much of the time between Lae in New Guinea and Howland, and for at least 1800 miles she would have no visual checkpoint. The only possible points of contact would be the USS *Ontario*, which was to be approximately halfway, the coastguard cutter *Itasca*, standing off the island, and the USS *Swan* between Howland and Hawaii: considerably less back-up than on the first attempt, timed to coincide with a Pan-American survey flight.

In Miami, Amelia impressed the mechanics with her willingness to 'work like a greased monkey when necessary' and eat with them at a 'greasy spoon' before she and Fred Noonan took off on 1 June. She posted log-books and articles from various stopping places: she was 'very glad' to see Fort Alexa in Brazil 'sitting just where it should be', commented that from the air the Red Sea was as blue as any other, and described how 'the water would have drowned us if our cockpit hadn't been secure' in a monsoon rain storm between Calcutta and Akyab in Burma. In Karachi, she telephoned George, 8000 miles away in New York, telling him she felt 'swell' and would post the stamp collectors' covers in Darwin or Lae.

On 30 June Amelia completed an article in Lae, and wrote of

rolling down the runway bound for points east: Howland, Honolulu, home by 4 July. In forty days, she and Fred Noonan had flown 22,000 miles, crossed the Equator three times, landed twenty-two times: she called it a 'leisurely trip'. The take-off from Lae was described over the American radio news service as 'a thrilling one, with the large plane getting off the ground with only fifty yards of runway to spare. Even now her huge plane is soaring over the South Pacific . . . ': at about the same time reports of her disappearance somewhere near Howland were beginning to come through.

The last message to be clearly heard sounded anxious and uncertain: 'We are on a line of position 157 dash 337. Will repeat this message on 6210 kilocycles. We are running north and south.' It was not repeated. The *Itasca*, off Howland, had picked up another message six hours earlier saying through static that the conditions were 'cloudy and overcast'. Neither lasted long enough for a bearing to be taken.

The ensuing search of 25,000 square miles of ocean, authorised by President Roosevelt, lasted a fortnight and cost an estimated $4 million. George Putnam sent lengthy telegrams giving unhelpful suggestions about how and where to search, including one based on Jacqueline Cochran's extrasensory powers. The aircraft was, she said, still floating; both Amelia and Fred Noonan were alive, although Fred was unconscious with a fractured skull. She mentioned the *Itasca* and a Japanese fishing boat by name and gave a fairly precise position, following the course of her friend's drifting for two days until Jacqueline sensed that Amelia was dead.

The aircraft carrier *Lexington*, carrying 300 naval aviators and 60 aircraft, arrived a week later. Paul Mantz rated the chances of Amelia having landed the aircraft on the sea and kept it afloat at one in a thousand. The Lockheed technicians said that if Amelia had been forced down by running out of fuel, the Electra's empty tanks would keep her afloat for no more than nine hours: once down, her radio would be useless.

Nothing was found. Criticism was levelled at Amelia's apparent carelessness. She had left her flares in the Burbank hangar, and had removed some radio equipment, lessening the range of communication. The aircraft carried no emergency portable radio. The official report stated: 'It is very apparent that the weak link in the combination was the crew's lack of expert knowledge of radio.' Neither Amelia nor the *Itasca* was fully informed about the radio frequencies available to the other, and no one had told Amelia about a high-frequency direction-finder on Howland.

An explanation was needed for her death. How could the woman who had seemed immune to danger and had so fervently campaigned about the safety of flying have died in one of the most expensive and best-equipped aircraft ever to be privately owned? The answer for many people, including her mother, was that she had not. Rumours were eagerly repeated: Amelia was, for instance, on a secret government mission to spy on the Japanese islands, although a confirmed pacifist with no specialist training and only a small tourist camera was an unlikely choice for a spying mission. The various versions of this story were equally unlikely, although it was true that the President had given priority to building the runway at Howland when there were many more pressing projects on the mainland. None of the so-called eyewitness accounts of Amelia's capture, imprisonment, execution, or continued existence have ever stood up to close examination. George Putnam was certain enough that she was no longer alive to remarry eighteen months after her disappearance.

To her many admirers, Amelia Earhart was the symbolic emancipated woman, the equal in the air and on the ground of any man, a winged legend. Her career in aviation helped to extend its possibilities, and those of other women, and her determination to lead an independent life was interpreted as a fight for women's rights. Her critics considered her not a very good pilot, found her over-confident and arrogant, and accused her and her husband of using friends in high places to get their own way. Amelia would no doubt have repeated that she flew for fun, was fortunate to be able to make her flying pay, and hoped that it might serve some practical purpose for the future of aviation, in which she had a deep and lasting conviction.

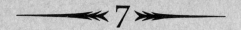

Amy Johnson:
British Heroine from Hull

Amy Johnson might be described as the British answer to Amelia Earhart. Unstable herself, and in a constant state of tension, often near or at emotional breaking point, she admired and envied Amelia's self-controlled success.

Amy, who was born in 1903, started her life in a street of terraced houses where nearly everyone was connected in some way with the sea. Her father, a prosperous Hull fish merchant, had taken over his family's business after unsuccessfully seeking his fortune in the Klondike gold rush before his marriage; during Amy's childhood, he gradually acquired the social status desired by his wife, who came from a well-established and respected but impoverished local family. The Johnsons' three daughters were brought up in a conventional Methodist atmosphere, but Amy, the oldest, was often moody and rebellious. She ascribed her later introspective moodiness to self-consciousness about the false replacements for her front teeth, which were broken by a cricket ball when she was fourteen. Because of her apparent arrogance, combined with a lack of social self-confidence, she had few friends. Her muddled dreams of romantic bliss and great achievements were nurtured by an avid devotion to the cinema, where she sat spellbound by heroines of the silent screen such as Pearl White, whose lurid adventures were accompanied by a wheezy piano.

By the time Amy went to Sheffield University, where her three-year degree course was to be followed by a year's teacher training, she had fallen in love. Although she and the object of her devotion, who was eight years her senior, became lovers, Amy took the affair considerably more seriously than he did: he was Swiss, and accepted her initial approach, her devoted company and her frequent long letters chiefly as a way of improving his English. For years her life was based on the conviction that they would marry, but although he took her to Switzerland to meet his mother, her lover had no such intention. She used the threat of emigrating to America in an attempt to force his hand and even offered to become, like him, a Roman Catholic. Her parents banned the relationship because Catholicism was abhorrent to them, but it nevertheless continued in secret. On several occasions Amy thought she might be pregnant.

When Amy had unenthusiastically completed her initial three years at university, she was too restless and dissatisfied to continue training as a teacher. Instead, she embarked on a commercial course with the intention of pursuing a business career, but gave up before acquiring more than mediocre secretarial ability. Her attempts to find employment were no more successful than her efforts to marry her lover: vaunting her degree, but with no professional qualification other than her incomplete skills as a secretary, she was rejected by one potential employer after another. Her first job, as a £1 a week shorthand typist in an accountant's office, ended with a nervous breakdown after a few months of boredom, misery and isolation from the other girls, to whom she felt herself superior. With a conviction that her abilities would soon lead to better things, she took a similar job with a Hull advertising company but again was unpopular with her female colleagues. Although she failed to persuade London advertising firms of her latent talents, she was certain that there she would find fame and fortune. Instead, she was taken on as a 'learner' at an Oxford Street store, Peter Jones, where she was soon in debt: she had not read the small print of her contract, and although her pay was to be £3 a week, until she had finished learning how to sell she was paid only on results.

Her next attempt at serious employment was more successful. As a copy typist in a solicitor's office, she showed enough aptitude and interest to be promoted to a senior partner's personal secretary, and was encouraged to study Company Law. With domestic bliss still in mind, she also went to cookery evening classes but, at last, reluctantly accepting that her lover did not intend to marry her, she decided in 1928, out of bravado, to take up flying. When the announcement did not bring him running back to stop her, she was disappointed. To prove she was serious, however, she wrote to the de Havilland School of Flying at Stag Lane on the northern outskirts of London, but could not afford its charge of £5 an hour for tuition. She had to make do with the vicarious excitement of such Hollywood films as *Wings*.

Later in the year, she discovered the London Aeroplane Club, also at Stag Lane: membership cost an initial three guineas, with an annual subscription of the same amount. As the club received government sponsorship, the cost of flying was only £1 10s an hour for tuition, going down to £1 a flying hour for members with a pilot's licence. On her salary of £5 a week, this outlay was just possible: there was, however, a waiting list for instruction. Amy was impatient, and attempted, unsuccessfully, both to jump the queue by writing to Lieutenant-Commander Perrin, secretary of the London Aeroplane Club and the Royal Aero Club, and to gain a flying scholarship. While she waited, as an Associate Member of the club, she gave her secretarial services free in the evenings to the Air League, an organisation devoted to persuading more people to become pilots so that Britain could keep up with Germany: although officially banned from having an Air Force, Germany had its own Air League with one million members, whereas Britain's had only 6000 members throughout the whole of the Empire.

When her lover married someone else, Amy's initial reaction was that flying would provide a satisfactorily dramatic way of committing suicide. After her first flying lesson in a club Moth her instructor told her she would never make a pilot. Even with a more sympathetic instructor, she showed little natural aptitude: she was too heavy-handed with the delicately balanced controls of the Moth, and made her first solo flight only after nearly 16 hours of dual flying. Generally a natural pilot would go solo after about 8 hours, and anyone who was not ready to do so after 16 hours was expected to give up trying.

In her free time, Amy enjoyed the social side of the club: normally withdrawn and ill-at-ease in company, and self-conscious about both her teeth and her Yorkshire accent, she relaxed in the bantering atmosphere, and made a nuisance of herself in the hangar. Women were not expected to be interested in the engineering aspects of flying – Lady Heath was the only woman who had ever qualified as a ground engineer, and she had done so in America – but the chief engineer at Stag Lane, Jack

Humphreys, soon recognised that Amy had the instincts of an excellent engineer. Under his protection she was initiated into aircraft maintenance and adopted the nickname of 'Johnnie'.

Amy gained her A licence in July 1929, and lost her way on her first cross-country solo flight. In December she became the first woman to qualify in Britain as a ground engineer. Determined to have a career in aviation, Amy asked her father for financial support so that she could give up her job: her boss was no longer prepared to allow his requirements to take second place to flying. Her outwardly vivacious married sister Irene, who had been nervous and delicate as a child, had just committed suicide, and Will Johnson was not initially prepared to support Amy in such a dangerous activity; but Jack Humphreys managed to persuade him to change his mind.

Amy was banking on a job as demonstrator of a revolutionary aircraft designed by a young man called James Martin, later to become Sir James Martin of the Martin Baker Aircraft Company and the inventor of the ejector seat. When she hinted to a London *Evening News* reporter, early in 1930, that she was planning a long-distance flight, she gained her first taste of publicity:

GIRL TO FLY ALONE TO AUSTRALIA

THE FIRST WOMAN AIR ENGINEER AND HER PLANS

A SECRET PLANE

Amy was delighted, in spite of the inaccuracies. Her age was given as twenty-two, instead of twenty-six; she did not come from the Midlands, nor had 'her skill with aircraft and engines' ever earned anything, let alone 'a comfortable living'.

After 50 hours of solo flying, Amy obtained her B licence and was given some maintenance work on a privately owned Moth: there were only about 60 aircraft in private ownership in the country, and more than half of them were de Havilland Moths. James Martin's 'secret' plane was nowhere near ready, but Amy hoped to find a sponsor and an aircraft for an Australian flight. Her appeals to Lords Rothermere and Beaverbrook and Sir Thomas Polson for at least £1000, preferably £1500, merely brought good wishes. As a last hope, she wrote to Sefton Brancker, Director of Civil Aviation, who made inquiries about the optimistic young woman who had forgotten to sign her letter. He discovered that the London Aeroplane Club considered her both a reliable pilot and an efficient ground engineer, and put her in touch with Lord Wakefield. Her father offered to pay part of the cost of an aircraft and Lord Wakefield agreed to provide £500: Will Johnson assured him that he would be 'helping a straightforward, good and clever girl to obtain her ambition and at the same time forward aviation'.

Less than three weeks before she intended to set out, Amy acquired a two-year-old Moth which already had extra fuel tanks, giving it a range of 13 hours flying. She christened it *Jason*, after the legendary Greek whose symbol was the trademark of the Johnson family fish business, had it painted bottle green with silver lettering, and set about her final preparations with assistance from its previous owner, Captain Wally Hope. Vaguely anticipating 'thrills and adventures', at Captain Hope's suggestion she equipped herself with a revolver, a mosquito net and a sun helmet for self-protection; a portable stove, reserve provisions, medicines and a first-aid kit for emergencies; and a parachute and fire-extinguisher. She ordered a set of maps fixed on rollers to cover her entire route, although much of it had never been surveyed in detail. With Sir Sefton Brancker's help, she managed to cut the usual time needed to obtain permissions to land in a dozen countries and fly over many more. She also assured her mother that she was 'taking every precaution'.

When Amy Johnson left Croydon on 5 May 1930, the longest flight she had made was 147 miles from Stag Lane to Hull. Her total solo flying time was only 75 hours, she had never flown with a full load of petrol or over the sea, and had rarely made a good landing. Her confidence was based on lack of experience: 'The prospect did not frighten me, because I was so appallingly ignorant that I never realised in the least what I had taken on.' The fumes in the cockpit whenever she had to pump petrol from one tank to another made her feel so sick that it was only the ignominy of giving up the flight which prevented her from doing so in its early stages: she had to pump 50 gallons of fuel every hour, each gallon taking forty strenuous strokes.

Her aim was to beat the time of the Australian pilot Bert Hinkler, who in 1928 had taken 15 days for the first solo flight from England to Australia. The first three days, although exhausting, went well. On the fourth, she flew into a desert sandstorm and her aircraft was buffeted so violently that she was forced to land. Sand and dust covered her goggles, her eyes smarted, and she had never been so frightened in her life. She put all her luggage as a break behind the aircraft's wheels, struggled to cover the engine and the air vent in the petrol tank to keep the sand dust out, and sat for three hours on *Jason*'s tail, her revolver ready for the ferocious desert dogs she could hear barking. When she eventually landed at Baghdad one of the wheel struts broke, strained by her high-speed desert landing: mechanics worked all night to repair it.

Her next stop was at the eastern end of the Persian Gulf, halfway between Baghdad and Karachi: the aerodrome had not been used for some time and, as she landed on the rough ground, coming in as usual too fast, a bolt securing the top of the new strut sheered. After gratefully sleeping for two hours at the house of the British Consul and his wife, she found to her delight that the bolt had been replaced, and carried out her routine mainte-nance work by moonlight. She reached Karachi two days ahead of Hinkler's time: he had taken eight days, averaging less than 600 miles a day; she had averaged 700 and reached Karachi on the sixth day. Her aim of beating his record, and her success over the first half of the flight, brought her as much praise and attention as she could have wished for.

When she realised that she would not have enough petrol to complete the 1000 miles over desert which separated Karachi from Allahabad, she landed on a parade ground at Jansi, twisting her way round trees and a telegraph post, scattering soldiers in her path, before coming to rest wedged between two barrack buildings. One of the officers described her as 'a girl – young, almost a child, wearing only a shirt, an ill-fitting pair of khaki shorts, socks and shoes, and a flying helmet. The skin on her face, arms and legs was burnt and blistered by the sun, and tears were not far from her tired eyes.' The damage to her aircraft was patched up, and she spent a restless, short and oppressively hot night, disturbed by dreams of impending crashes.

Between Calcutta and Rangoon, she flew through monsoon rain. When she saw crowds of people waving at her from a balcony she assumed that they were welcoming her to the race course where she intended to land. She realised her error too late as goal posts and a barrier of trees made retreat impossible, and tipped into a ditch. The teachers and pupils of the Government Technical Institute five miles north of the town had not been waving but pointing the way. With their help, she repaired the damage. The propeller was replaced by the spare she was carrying with her; the broken wing ribs were painstakingly copied in the school workshop and covered with material from men's shirts made of war surplus aeroplane fabric, which was stiffened and waterproofed with 'dope' made up by a local chemist. Still in pouring rain, *Jason* was towed by the local fire engine to the race course.

Amy felt her way over the mountains to Bangkok, where she spent the night at the aerodrome. On the way on to Singapore, she took off her goggles because she could not see through them in the rain, and found herself circling a field instead of following the coastline. When the weather at last cleared, she decided to make an unscheduled stop at Singora. Next day at Singapore several hundred Europeans greeted the new flying heroine, whose face was burnt and smeared with oil and whose ill-fitting man's shorts and heavy socks, oil-stained brogues, drill jacket and topee contrasted with their garden party clothes.

Bad weather and yet more crash landings and delays for repairs put Hinkler's record out of reach. Miraculous escape followed miraculous escape, and she had good reason to feel that some guardian angel was watching over her. Over the Java Sea, the clouds parted in the nick of time – 'a happy manifestation'; she ripped the fabric of *Jason*'s lower wings on bamboo stakes when she made an emergency landing on Java; after another unplanned stop, near a remote village, she was reported missing; over the Timor Sea, she recited nursery rhymes to keep herself calm in bad visibility and with a spluttering engine.

When she reached Darwin, on 24 May 1930, Amy Johnson felt that she had failed: she had taken 19½ days, four days longer than Hinkler. The Australian and British publics thought otherwise: the first woman to fly solo to Australia was an international heroine. She received congratulatory telegrams from King George V and Queen Mary, Ramsay MacDonald, the prime minister, the King and Queen of Belgium, the Lindberghs, Louis Blériot, Francis Chichester, and hundreds of other people, and was mobbed and fêted wherever she went. Then after one night's rest, she was plunged into an aerial tour of Australia, impressing everyone who met or saw her with her naturalness and modesty.

Her composure, however, was a façade. She became increasingly exhausted and tense, often weeping hysterically as soon as she was out of the public gaze. She had wanted publicity, but had imagined nothing like the ordeal of being constantly adored by thousands. On the fourth day of the Australian tour, she crash-landed at Brisbane, and for the rest of the tour travelled as a passenger. She realised that her original intention of flying herself back to England was more than she could cope with, and was near to a complete breakdown by the time she started her return journey by commercial airline.

When the *Daily Mail* announced that it was awarding Amy Johnson £10,000, 'the largest amount ever paid for a feat of daring', a Sydney scandal sheet, *Smith's Weekly*, called her 'The Air Digger of the Sky' and 'The Gimme Gimme Girl'. In England, a eulogistic book *Amy Johnson: Lone Girl Flyer* had been rushed into print by a journalist called Charles Dixon whom she had never met, and Amy found herself committed to a gruelling tour of forty towns organised by the *Daily Mail*.

After the first week she reached the point of collapse and the tour was cancelled, although she was allowed to keep her £10,000. She had been given two brand-new aircraft – a de Havilland Puss Moth, *Jason II*, and a new Gipsy Moth, *Jason III*, as well as an MG car. The first *Jason* has been in the Science Museum in London ever since, its smallness and frailty a tribute to the courage, endurance and foolhardiness of its second owner whose name was on everyone's lips in a spate of popular songs: 'Johnnie, Heroine of the Air', 'Queen of the Air', 'Aeroplane Girl', 'The Lone Dove', and, most popular of all, 'Amy', which had the refrain 'Amy, Wonderful Amy'.

Amy did not feel wonderful: her dream had become a nightmare, from which she tried to escape on another long-distance flight: across Siberia to China. It was hardly a sensible route to choose in winter and, after struggling across Europe in January 1931 in icy fog and snow, she abandoned the venture when she crash-landed in a potato field near Warsaw. In the summer she set out again with Jack Humphreys: but their ten-day light aircraft record for the 7000 mile flight to Tokyo was overshadowed by a nine-day solo flight from Australia to England, also in a Puss Moth, by an unknown pilot called Jim Mollison.

Amy's plans for a round-the-world flight with Jack Humphreys, and a lecture tour – 'How *Jason* and I flew to the Land of the Golden Fleece' – came to a halt when she collapsed with acute

abdominal pain: officially, she had an appendectomy, probably
a euphemism for a hysterectomy. She took a prolonged sea
cruise to recuperate, and in Cape Town met Jim Mollison when
he arrived at the end of a five-day record-breaking flight from
England. Two years younger than Amy – he was twenty-seven –
he seems to have been, although on the short side, irresistibly
attractive to women, but the rumours in the press of a romance
between the two pilots were premature. Amy continued her
voyage to Durban suffering from yet another nervous collapse.

Not long after her return to England, however, she had lunch
with Jim Mollison at a smart London restaurant: over the brandy,
he proposed and she accepted. They had met for only a few
hours, and although they shared a major preoccupation with
flying they were, as they might have discovered had they both
been less impetuous, entirely unsuited to each other. Amy, for all
her new sophistication – she dressed in the latest fashions, took
trouble with her hairstyle and make-up, tried to lose her Yorkshire accent – was by nature puritanical; Jim was irresponsible, a
heavy drinker, and a womaniser. Amy was plunged into a social
whirl of parties, publicity and wild spending. Her family was not
invited to her wedding in July 1932.

A few days later Jim Mollison set off from Ireland on a multi-
record-breaking fight across the Atlantic from east to west. Amy
retaliated by beating his London to Cape Town solo record by 11
hours and, by making a record flight back, gained the record for
the round trip as well. The *Sunday Dispatch* serialised her life
story, and she was awarded the Segrave trophy as the person
who, during 1932, had given 'the most outstanding demonstration of the possibilities of transport by land, air or water'. The
following year, Jim won the Royal Aero Club's Britannia Trophy
for 'the most meritorious performance in the air during 1933' for a
flight across the South Atlantic.

For the Mollisons' first joint flying venture, they planned to fly
to New York, from there to Baghdad, and then back to London,
and to create a world long-distance record on the second leg with
a twin-engined de Havilland Dragon in which three extra fuel
tanks replaced the usual passenger seats: but they crashed on
take-off. A month later, in July 1933, they took off successfully
from a Welsh beach. Fifty miles from New York – Amy had
wanted to land at Boston to refuel, but had been overruled – they

ran out of petrol over the airfield at Bridgeport, Connecticut, and overshot the runway in the dark. The flight ended with both of them in hospital: Amy was bruised and shaken, and Jim, who had been knocked out, had gashes on his head and face. This ensured more than adequate publicity, but it was Amy who became the darling of the New York press: Jim, who was intensely competitive, resented both his wife's popularity and her new friendship with Amelia Earhart, who invited them to stay during their convalescence.

The American pilot's successful career in aviation was just what Amy had hoped for for herself, and when Jim returned to England she stayed in America. A month later, Jim returned, with a new aircraft provided by Lord Wakefield, but their record attempt was abandoned when, after waiting for several weeks for fair weather, they were unable to lift the aircraft with its heavy load of petrol off three miles of hard sand beside a Canadian lake. Amy's health was again causing problems, and she went into hospital in New York for a month with nervous exhaustion and an internal ulcer: she was ordered to take at least six months' rest. When she returned to England, she was appointed aviation editor for the *Daily Mail*; but she found little pleasure in writing, although she had an inflated idea of her ability.

In the autumn of 1934 the Mollisons entered the Great Australian Air Race. Their preparations were made hastily: a fortnight before the start, they had still never flown the de Havilland Comet in which they were to compete, and had neither maps nor equipment. They nevertheless halved the London-Karachi non-stop record on the first leg, but lost their way and then had to retire when one of their engines seized up on the second stretch. Amy returned to England alone on an Imperial Airways flight.

In the little time that the Mollisons spent together, Jim drank heavily, and Amy soon followed suit to deaden her disgust. Alcohol did not combine well with her emotional instability, and when she was stopped for speeding after a series of parties in Miami she reacted by hitting one of the traffic police, the first of a number of motoring contretemps in both America and England. In January 1935 Jim left for America without telling Amy he was going. She admitted to her father that she was 'still very much in love with Jim . . . but . . . as unhappy with him as away from him'. Unable to tolerate his fequent amorous adventures and heavy drinking, Amy moved to Paris.

With the sponsorship of a wealthy French businessman, she formed a company called Air Cruises: in co-operation with luxury hotels she intended to be the pilot on pleasure trips round Europe. To coincide with the start of Air Cruises, and to escape the amorous advances of her backer, Amy set out on another record attempt to Cape Town. Dressed as for 'a normal business trip' – in a Schiaparelli suit and matching coat – to underline the comfort and safety of flying, she crashed her elegant pale blue Percival Gull on take-off from Colom Bechar in Algeria. Such setbacks always brought out her most determined qualities, and a month later she set out again, reclaiming the record, recently taken from her by Tommy Rose, by 11 hours on the way out and setting further records for the return flight and the round trip.

Once again Amy Johnson was 'wonderful': 'Her last flight', according to *The Times*, 'is no mere flash in the pan but an achievement in keeping with a distinguished aeronautical career.' Her private life, however, was far from satisfactory, and she came to the reluctant conclusion that Jim was incapable of remaining faithful to her. In October 1936 she broke her nose and dislocated her shoulder when she crashed her imported American Beechcraft, not for the first time, and simultaneously announced that she and Jim were separating. After another nervous breakdown, cured by a skiing holiday, she planned to take part in a transatlantic race, which was cancelled, and fell out with her French backer and Air Cruises. When Amelia Earhart was lost over the Pacific, Amy assured her mother that there would be 'No more flights so no need to worry!'

In 1937, under her maiden name, Amy found that being famous was still no help when it came to serious employment in aviation. She took up gliding, a sport recently introduced to England, and rediscovered in the gliding community the friendliness, enthusiasm and sporting spirit which she had enjoyed so much in her early flying days. The slowness and tranquillity of gliding was a welcome relief from the 'deafening roar' which had accompanied her for so many years. In 1938, she divorced Jim Mollison. Financially and emotionally she was back where she had started.

In the summer of 1939 Amy Johnson found a job as a pilot for a Solent ferry company – Portsmouth, Southsea & Isle of Wight

Aviation Ltd – flying on nocturnal army co-operation contracts to provide moving aerial targets for the searchlights and range-finders of anti-aircraft batteries. In her spare time she discovered the pleasures of sailing – 'much more exciting than flying' – and even planned to sail round the world. At last her life seemed to have settled into a reasonable and enjoyable pattern, but after only a few months it was interrupted by the outbreak of war. When, as probably the most experienced woman pilot in England, Amy was invited to join the first women's contingent of ferry pilots with the new Air Transport Auxiliary (ATA), she refused: she felt that her experience and reputation fitted her for something more distinguished.

When the Solent ferry firm was taken over by the Royal Air Force, she reluctantly applied for a £6-a-week job with the ATA, where at first her attitude showed the arrogance which had made her unpopular with her colleagues in her pre-flying jobs. She was offended at being summoned for an interview like any 'typical CAG (Civil Air Guard) Lyons-waitress-type', resented having to conform as a 'new girl', and hated flying slow open aircraft. 'The whole of my flying career has just been one long fight against jealousy,' she complained to her parents, 'and I'm right up against it here.' But gradually her hurt pride gave way to positive enjoyment, and wartime colleagues remember her as 'just mucking in with the rest of us'.

When her ex-husband also joined the ATA, the *Sunday Express* hinted at a fairy-tale ending: a figment of the reporter's imagination. Amy seemed incapable of forming a stable romantic relationship, although according to one ATA colleague she was 'a very sexy person' and 'chased after stupid little squidgies who treated her cavalierly'. She was nevertheless happier than at any time since her childhood, although she was incensed that she was paid less than a man doing the same job. It was a cruel irony that, after so many years of unhappiness and dissatisfaction, she should be the first member of the ATA to be killed.

She was thirty-seven and had accumulated more than 3000 hours in the air when, in January 1941, she again made headlines: 'AMY JOHNSON BALES OUT, MISSING'. She had been on a routine flight, delivering an aircraft from Prestwick to Kidlington, and had made an overnight stop because of bad weather at Blackpool. Although her route should have taken her well inland, she died in the Thames estuary. Her death, like that of Amelia Earhart, was surrounded by mystery: eyewitness reports agreed that her aircraft plunged into the sea, and that at least one person baled out, but there were conflicting statements about a second person in the water. The only person who might have known whether, as was rumoured, Amy Johnson was carrying a secret passenger, was a naval officer from HMS *Haslemere* who perished in an attempt to rescue her. Her body was never recovered, although her flying bag was picked up, the only clue to the identity of the pilot who had disappeared under the water.

Weather conditions were such that a more cautious pilot might not have taken off: but there was an ATA party which Amy was anxious to attend, and she was always concerned about losing face. Although there was an official ATA ruling that pilots should not fly above cloud in the hope of finding a gap in time to land, Amy was by no means the only pilot to ignore the ruling. Having 'gone over the top', she had probably flown off course in the murk until she ran out of petrol.

Pauline Gower, the ATA Commanding Officer, was loyal in her protestation that Amy would never have disobeyed regulations by picking up a passenger without official clearance – a crime not entirely unknown in the ATA, although it was strictly forbidden – nor would she have flown unwisely or dangerously: but Amy Johnson had in the past often enough acted irresponsibly and flown dangerously. The rumour of some secret mission, however, seems to be more the sort of wishful thinking in which she might herself have indulged than a serious likelihood: as a fellow ATA pilot put it, 'Can you imagine someone in MI5 saying "Let's send one of those women in their bright yellow trainers" when they had plenty of well-qualified men pilots available?'

Amy Johnson's life was unsettled and often unhappy. Fame, and the brief fortune which accompanied it, brought her little happiness. Reality always proved far tougher than her romantic dreams, but on her long-distance flights she found reserves of strength which, combined with an unusual amount of sheer luck, kept her going. To recognise her weaknesses is not to denigrate her achievements: that she overcame her limitations as a pilot and as a person to become 'Amy, Wonderful Amy' to millions of people is a great achievement in itself.

8

Anne Morrow Lindbergh:
A Hero's Wife

Charles Lindbergh was not at all the sort of man Anne Morrow had expected to marry, and she certainly never anticipated that marriage would involve her in long-distance flying with a man considered by millions to be a hero. She met him in 1927 when he was at the height of his fame, a few months after he made the first solo crossing of the Atlantic and won the unwanted adulation of millions as well as the $25,000 Orteig Prize.

In 1919, Raymond Orteig, a Frenchman who owned hotels in New York, had offered $25,000 to the first pilot or pilots to fly non-stop across the Atlantic between New York and France. Alcock and Whitten-Brown had made the first Atlantic crossing that year, but by a much shorter route, and no one had claimed the reward by 1926, when the offer was renewed. The first four men to take up the challenge, in a specially designed Sikorsky, crashed on take-off: two of them died in the resulting fire. Others, including Commander Richard Byrd, announced their intention to try for the prize, among them an impecunious and unknown twenty-five-year-old mail pilot, Charles Augustus Lindbergh, an ex-barnstormer and stunt parachutist.

Backed by a group of St Louis businessmen, Lindbergh intended to fly solo in a small Ryan monoplane, *Spirit of St Louis*, but at first received no attention as a serious contender. When he succeeded in the first ever solo Atlantic crossing, in 33½ hours, he also achieved the longest flight ever made.

A puritanical young man with a deep belief in the future of aviation, a strict regard for truth, and an aversion to alcohol and exaggeration, Lindbergh loathed the publicity which engulfed his life. Inaccurate press reports incensed him, and reporters and photographers irritated him with their stupidity and importunity, although he was pleased enough that his fame was accompanied by financial rewards great enough to solve his many money problems. Now the USA's most eligible bachelor, Lindbergh was tall and slim, with a grin which delighted his admirers, but he was awkward and ill-at-ease with women and embarrassed by their devotion.

He was more impressed with the attention paid to him by the millionaire aviation enthusiast Harry Guggenheim, who invited him to stay at his baronial-style mansion and ran the Foundation for Aeronautical Research. At the Guggenheim residence, shielded from sightseers by a 350-acre estate with a private airfield, Lindbergh met only people selected by his host for his social and professional advancement: Juan Trippe, president of Pan-American Airways, the aviation pioneer Orville Wright, the publisher George Palmer Putnam, politicians and financiers. Among these was Dwight Morrow, lawyer, international banker and at the time the American Ambassador in Mexico. Lindbergh, America's charismatic hero, seemed the ideal person to foster goodwill between the two countries.

By the time Anne Morrow joined her family for Christmas, Lindbergh had made the first non-stop flight from New York to Mexico, which seemed to be under his spell: so, it appeared, was the Morrow household, where he was a guest. Anne, who thought herself immune from hero worship, resented his intrusion into their closely knit family circle and was convinced that she could neither like nor have anything in common with him. She was, however, fascinated by the contrast between his self-assured public image and his apparent shyness, and poured out her attempts to analyse his attraction in her diary, where she also confided her regret that she was too young and silly to interest such a superior being.

Anne Morrow at twenty-one was shy, introspective and self-consciously aware of her own imagined inadequacies. Her interests were literary – she wrote poetry, won essay prizes, read widely and searched her soul regularly in her diary. Although she felt herself to be Charles Lindbergh's intellectual superior, she longed to share and understand his feelings and experiences. When he took her and her sisters for their first flight, the combination of his presence and her 'REAL and intense CON-

SCIOUSNESS of flying' created what she called a complete experience. For several months Anne's diary and her letters to her younger sister Constance were monopolised by the enigma of the flying hero. The two girls even went flying together secretly, and imagined a wedding between Lindbergh and their older sister Elizabeth. Anne was 'sick with amazement' when she saw the film *Forty Thousand Miles with Colonel Lindbergh*, and asked herself 'Did I ever meet this boy? How could I possibly comprehend him or it?'

She was so lacking in self-confidence that when, nearly a year after their meeting in Mexico, Charles Lindbergh telephoned her, she was tongue-tied and could only apologise that there was no one else at home. She nevertheless accepted his invitation to go for a flight, for which she fitted herself out in her sister's riding trousers, her mother's woollen shirt and blue burberry, her father's thick grey golf stockings, and carried a leather coat: she had not realised that they were to have lunch at the palatial home of Harry Guggenheim.

Anne soon accepted that Charles Lindbergh sought and enjoyed her company as much as she wished for his, but although they attempted to avoid publicity it was impossible to maintain secrecy about their meetings for very long. Before the year ended she wrote to a friend: 'Apparently I am going to marry Charles Lindbergh . . . there he is, and I've got to go . . . Wish me courage and strength and a sense of humour.' The Charles Lindbergh she was marrying was not the Lindbergh of the newspaper acounts, but a young man with 'vision and a sense of humour and extraordinarily nice eyes'. Their engagement was formally announced in February 1929, and Anne was congratulated on her good fortune as bride-to-be of the 'Prince of the air'. The secrecy of the wedding at the Morrow home in Engelwood in May 1929 antagonised the press, and during the young couple's honeymoon on a boat they were hounded by reporters and photographers. Charles interpreted the interest shown in his marriage as an offensive intrusion of privacy, and Anne was required to be as secretive as he was, smothering her natural spontaneity in case she said or did anything which might be reported. She soon dreaded 'the leer of recognition', and felt free only when she was flying with her husband.

Charles Lindbergh insisted that his wife must be a useful crew member: she learned morse so that she could be his radio operator, how to read a chart so that she could be his navigator, and had to be able to take over the controls in an emergency or to give him a rest. Fortunately her enthusiasm for flying was almost as great as his own. They had been married for less than a month when they inaugurated a Transcontinental Air Transport (TAT) route across America from New York to Los Angeles. Anne was the only woman on the first flight with paying passengers, from Los Angeles to Winslow, in a tri-motor Ford with luxurious reclining leather-covered chairs, small individual folding tables, and the attention of a courier who brought in-flight refreshments: soup in the morning, tea in the afternoon.

In September, they flew to Albuquerque in a Lockheed Vega to investigate a TAT crash, then spent two weeks on Pan-American survey flights over the Antilles and in South and Central America. By the end of October, Anne was suffering almost constantly from nausea and sickness, and in November it was confirmed, although only privately, that she was pregnant: Charles was outraged when asked at a press conference whether she was expecting a baby. As soon as the sickness of early pregnancy abated they flew together to California to supervise the final construction of a new single-engine low-wing Lockheed Sirius monoplane with which they later surveyed Pacific and Atlantic air routes. In spite of her pregnancy, Anne took up the new aerial sport of gliding in California, and was the first woman in America to gain a glider pilot's licence.

Two months before the birth of Charles Lindbergh junior, Charles and Anne Lindbergh made a record flight from Los Angeles to New York. For four of the 14 hours of the flight, Anne was feeling far from well. The weather was bad, but to allay public fears about the safety of aviation – two commercial aircraft had recently crashed – they were determined to continue, flying at an altitude which, without oxygen, caused Anne considerable discomfort and sickness. The baby was born on 22 June 1930, Anne's twenty-fourth birthday, but the birth was not announced until two weeks later.

Soon after their son's first birthday, his parents left him in the care of a nanny for two months while they flew in their Lockheed Sirius on a new northern route to the Orient: to Japan via Alaska and Siberia. Anne's long and frequent letters home combined an intense interest in everything she saw and everyone she met, and the concern of a young mother separated for so long from her baby. She described places where no aircraft had ever been seen, and where, as no one had ever heard of the great Charles Lindbergh, they were rarely troubled by publicity.

She had recently passed her radio operator's exam and was in sole charge of the radio, gradually increasing in confidence and ability. Wearing flying suits with electrical heating, they crossed Canada and Alaska, flew along the shores of the Arctic Sea, across the Bering Sea to Siberia, down to Japan, and finished the tour in China. The Sirius, converted into a seaplane and with extra fuel tanks, had a range of 2000 miles, but had to be equipped for an emergency landing in arctic and tropical conditions, on land and at sea. Everything they carried with them was carefully weighed, and by the time they had included a 25-lb anchor, rope, a rubber boat, oars and sails, camping equipment and emergency food supplies, a medicine kit, firearms and ammunition, as well as the essentials needed for flying and repairing the aircraft, they were left with only an 18-lb personal baggage allowance each.

They stopped at Arctic settlements where supplies were brought by ship only once a year, including Barrow where a radio operator sent daily weather reports to the rest of the world, and were entertained with Eskimo sports and a wolf dance. In Kamatchka, further from Moscow than they were from San Francisco but nevertheless part of Russia, Anne talked about babies with an English-speaking woman zoologist. As they approached Japan, they were forced by thick fog to land on the open sea 200 yds off one of the islands linking it with the far eastern Russian peninsula, which is as long as England and Scotland together. A wave of fear 'like terrific pain' swept briefly over Anne, 'shriveling to blackened ashes the meaningless words "courage" – "pride" – "control" '. They slept in the baggage locker, safely at anchor in the lee of the island: during a second night at sea a tornado raged to the west of their anchorage.

When they reached Nanking in China, the Lindberghs found that the Yangtze river had broken its banks. A vast flat area was flooded: crops, livestock, houses, even whole villages, had been swept away and countless lives lost. There was no hope that the flood water would subside before spring, several months away, and many of the homeless were living on drifting flat-bottomed sampans. As there was no aircraft in China with the range needed to survey the flooded area, the Sirius was used. Anne piloted while Charles sketched and mapped: in a day, they covered an area the size of Massachusetts.

For several days, they were guests in Hankow on board the British aircraft-carrier *Hermes*. The current in the swollen river was so strong that, to ensure its safety, the captain arranged for the Sirius to be hoisted on board every night. When all but one area had been surveyed, the Sirius was caught by the current as it touched the water, swung round, and one wing went under. Charles and Anne were already in their separate cockpits, and had to jump for their lives as the aircraft turned over and they were thrown into the water. Anne, to her surprise, was not afraid: merely annoyed that, after brushing her teeth so carefully in boiled water, she had swallowed some of the polluted river. The Sirius was righted and hoisted back on deck, but had been badly damaged.

In February 1932, the Lindberghs, who had so far been based at the Morrows' residence, moved at last into their own home, Hopewell, a newly built ten-roomed house with a separate servants' wing for a married couple who took care of all domestic details. Their son, Charles junior, was looked after during the week, which they still spent with the Morrows, by an English nanny: at weekends, Anne looked after him herself at Hopewell. The site for the house had been chosen from the air for its isolation: although it was only twenty minutes' drive from Princeton, and an hour and a half to New York, it was approached by a long single-track drive and surrounded by 500 acres of bleak marsh and mountain countryside.

On the last weekend of February, both the parents and their son were suffering from colds, and decided to stay on at Hopewell, where they were joined on Tuesday 1 March by the nanny. When she put the child to bed, she bolted all the windows except one, where the shutters were warped. It was a stormy night, and no one heard the intruder who later put a ladder up to the window, climbed in, removed the sleeping child from his cot, and climbed out again.

For weeks, the parents and the police pursued every possible line of inquiry, including a number of cruel and time-wasting hoaxes, in the hope that Charles junior would be discovered alive. The press, although following the story avidly, showed more consideration for the grief of the parents than did the public. The Lindberghs received 38,000 letters: 12,000 people wrote about dreams; 11,000 expressed sympathy; 9000 wanted to make suggestions, and 500 claimed that the Lindberghs had got what their pride, arrogance and affluence deserved. After lengthy negotiations with a go-between, a ransom of $50,000 was paid to a man claiming to be in touch with the kidnappers.

More than two months after Charles Lindbergh junior was kidnapped, Charles Lindbergh senior was following yet another false trail, and was at sea searching for the boat where his baby was supposed to be being looked after, when the child's body was found buried in woodland not far from their home. He had died on the evening of the kidnapping from a blow on the head, probably while he was being carried down from his bedroom. Anne confided to her diary that it was a relief to know something definite, to be able to say 'then he was living', 'then he was dead'. The search for the killer involved her less than it did Charles: for him, it was essential that somebody should be brought to trial and punished, whereas Anne wrote that it was enough to know that he was dead, to come to terms with grief and to start a new life with the second baby she was carrying.

Jon Lindbergh was born in mid-August 1932, just over two months after his brother's body was found. It was October before his parents decided on a name, and Anne at first struggled to look on him as a new individual and not a repetition of her dead child. Charles Lindbergh appealed to the press to leave him and his family in peace, acquired an Alsatian guard dog, and presented the house at Hopewell to the state of New Jersey as a children's home: Anne could never feel happy there, and they returned to the security of her family home, where Charles often felt himself an intruder.

Before the new baby was a year old, his parents left him in the luxury of the Morrow house while they made another even more ambitious survey flight for Pan-American in the Sirius. They flew from America via Greenland, Iceland, Scandinavia and Russia to England, then continued south to Spain, West Africa, across the Atlantic to Brazil, over the Caribbean to Miami. Anne acted throughout as radio operator and navigator, and with her husband faced the dangers of ice floes, blizzards, sandstorms and tropical hurricanes.

On their return to New York after nearly six months, they hoped, in vain, to settle down to a peaceful family life with their son, and find a home where they could be together without the pressure of Anne's family. Instead, Charles Lindbergh was soon involved in a bitter battle with the Roosevelt administration over the United States airmail services. Allegations were made about improper business practices, and Lindbergh's own dealings with TAT were brought under suspicion. When Roosevelt cancelled the commercial airmail contracts, giving the work to the Army Air Corps, five pilots were killed, six injured, and eight aircraft wrecked, within a week: this was precisely what those who had resisted the move had feared, as the army pilots had inadequate experience of the rigorous flying conditions of the airmail service. Lindbergh, who had been one of the first airmail pilots, attacked the president in the press: using the press was one thing, being used by it another.

In September 1934, Bruno Hauptmann, a German carpenter from the Bronx, was arrested for the murder of Charles Lindbergh junior. The publicity which they dreaded again took possession of the Lindberghs. Harold Nicolson, who described Anne as 'tiny, shy, retreating . . . a tragedy at the corner of her mouth', was a guest at the Morrow house while he was working on a biography of Dwight Morrow, and encouraged her to continue work on her first book, *North to the Orient*. She made only two court appearances during the six-week trial at the beginning of 1935, but Charles sat in court throughout. When Hauptmann was condemned to death – although, as Ludovic Kennedy has argued, there has continued to be considerable doubt about his guilt – Anne wrote in her diary: 'The trial is over. We must start our life again, try to build it securely – C. and Jon and I.'

It was, however, impossible to lead an entirely normal life. There was little privacy in the Morrow household, and in spite of all their precautions Jon, who was driven to and from a nursery school every day by one of the teachers, was twice accosted by photographers. Early in December, a campaign was mounted to repeal the death sentence passed on Bruno Hauptmann: yet again there would be the nightmare of publicity. Charles had had enough, and told Anne to start packing. On 22 December 1934 they left by ship for England. The *New York Times* condemned American society for forcing its hero into exile:

> The departure of Colonel and Mrs Lindbergh for England, to find a tolerable home there in a safer and more civilised land than ours has shown itself to be is its own commentary upon the American social scene. Nations have exiled their heroes before; they have broken them with meanness. But when has a nation made life unbearable to one of its most distinguished men through a sheer inability to protect him from its criminals and lunatics and the vast vulgarity of its sensationalists, publicity-seekers, petty politicians and yellow newspapers?

The *Daily Mail* asked the British to 'LEAVE THEM ALONE', and they were undisturbed by unwanted publicity, according to Anne's diary, until their association with pre-war Germany caused offence. Between 1936 and 1939 they made several trips to Germany as the guests of the Nazi government and with the semi-official US brief of finding out about German air power. The Nazis showed them exactly what they wanted them to see: enough aircraft and factories to give a gross over-estimation of the country's airborne ability. Charles Lindbergh became one of Germany's most convinced apologists: he could not believe that he was being used, refused to accept that the Nazi government might be a force of evil and, since he also believed that no one could beat the Germans militarily, argued that the buffer of a strong Germany was the only defence against the far greater threat of communism.

In 1937, Anne spent the sixth to eighth months of her third pregnancy accompanying her husband on a strenuous flight to India and back. In the autumn, they went to Germany, and on a third visit the next year Charles was unexpectedly presented with a medal by Goering, which did nothing to improve his tarnished public image: if he accepted the award, he would cause offence to the British and Americans, but if he refused it, he would offend his German hosts. Flattered as well as embarrassed, he accepted. Anne, less politically naïve than her husband, called the medal the 'Albatross', but seriously considered living near Berlin. Instead, they took their children to France, where they spent several unusually happy months as the only inhabitants, apart from their small staff, on an island off the north Breton coast.

Even when war was declared, Charles Lindbergh, loyally supported by his wife, refused to change his opinion of Germany: he took his family back to America, where he became deeply involved in the anti-interventionist movement and made frequent speeches which sounded more and more like the rantings of the Nazis. The *New York Times* criticised one of his speeches in 1940 for its political naïveté, ending with the cutting comment 'Colonel Lindbergh remains a great flier'. In the same year, Anne had another baby, a girl, and published *Wave of the Future*, a muddled attempt to support her husband's stand based on vaguely pacifist ideas. Harold Ickes, Roosevelt's Secretary of State for the Interior, condemned it as 'the bible of every American Nazi, Fascist, Bundist and Appeaser'.

Because of their conflicting opinions, the Lindberghs moved out of the Morrows' house. Even when they were ostracised by their friends, Anne never wavered in her support of her husband, and felt a humble sense of gratitude that she was the wife of such a great man. When America eventually entered the Second World War, Charles Lindbergh offered his services to the USA Army Air Force: he was turned down. Instead he worked with Henry Ford to produce aircraft to hasten the end of the conflict, kept wisely out of the political arena and was gradually accepted again by his former friends and colleagues. In the last year of the war, he flew in the Pacific as a civilian adviser to the US Army Air Force, and when war ended was sent to Germany to recruit leading German aviation and scientific experts. His sympathies were still with the Germans, and even when confronted with the reality of a concentration camp, a place where he felt that 'men and life and death had reached the lowest form of degradation', he equated Nazi atrocities with the general brutalising effect of all war on all nations.

In the years after the war Charles Lindbergh was employed by the US Army Air Force as adviser on space and rocket research projects. The Lindberghs lived on the Connecticut coast, at Darien, where Anne was often abandoned with their children: the youngest, another girl, having been born in 1945. Although her life was full, with frequent foreign travel, she was restless and unhappy, and was deeply hurt by a review which called a collection of her poems 'an offensively bad book'. After her mother died, in 1955, Anne felt that she must have some time to come to terms with herself, alone: in the relaxing solitude of her retreat, a small simple house on a beach, she tried to sort out her feelings in a book, *Gift from the Sea*, which made her one of the most popular writers in America. She saw it as her duty to herself and her family to combine the supportive role of a wife and mother with her own desire to write. Charles was as pleased as she was when her books were praised, and often helped her with suggestions and proof reading.

During the 1960s, Charles Lindbergh, who was working on an American supersonic aircraft, became increasingly disillusioned with the way aviation was progressing and with the devastation to wild life caused by so-called progress. For the last few years of his life, with Anne's relieved support, he became deeply committed to the conservation efforts of the World Wildlife Fund. He and Anne found a new peace together on the Hawaiian Island of Maui, where they built a house as simple as the one to which Anne had retreated to write *Gift from the Sea*. It was there that Charles Lindbergh died, in 1974, at the age of seventy-two.

For forty-five years, he had influenced developments in aviation, from world-wide passenger travel to the exploration of space. Anne had always supported 'the big things in C's life', and had felt that, whatever her own private inclinations, it would have been 'just plain silly' not to accompany him on his pioneering flights: if she had not done so, there would have been 'not much point in being married'. On one of their first meetings, Charles Lindbergh had told Anne of his interest in 'breaking up the prejudices between nations, linking them up through aviation': it was an impressive aim, and one towards which – in spite of public pressure, private tragedy and an unfortunate digression into politics – they had worked together.

9

Jean Batten:
Daughter of the Sky

While Amy Johnson was searching for a sponsor for her Australia flight, the equally ambitious nineteen-year-old daughter of a New Zealand dentist was also planning to fly from London to Australia. Jean Batten had not yet learnt to fly, but knew exactly what she wanted: to combine her love of travel with the personal satisfaction of flying farther and faster than other people.

She was born in September 1909, a few weeks after Blériot's first cross-Channel flight. Even as a young child, first in Rotorua where geysers, boiling mud and thermal pools drew tourists from all over the world, and then in Auckland, Jean longed to travel. When she was ten, she shared the general excitement at two great flying achievements: Alcock and Whitten-Brown's flight across the Atlantic and the first flight from England to Australia, made by the brothers Ross and Ken Smith. Her parents hoped that she would become a concert pianist, but her favourite school subject was geography, and in 1928 she decided that she wanted to emulate her heroes by exploring the world by air: it was the year in which Hinkler made his solo flight from England to Australia, Charles Kingsford-Smith and Charles Ulm flew across the Pacific from America to Australia in the *Southern Cross*, and Kingsford-Smith then linked Australia and New Zealand by air across the Tasman Sea.

Jean met Kingsford-Smith during his subsequent tour of New Zealand, and while she was on holiday in Australia early the following year he took her for her first flight, over the Blue Mountains. She felt 'completely at home in the air' and decided that, as England seemed to be the world's aviation centre, she would embark on her flying career there, accompanied by her mother. Her piano was sold to raise money for her tuition, and not long after her arrival Jean had her first flying lesson at the London Aeroplane Club, where she met other women who had learnt to fly at Stag Lane, among them Lady Heath, Lady Bailey, the Hon. Mrs Victor Bruce, and Amy Johnson.

No sooner had Jean Batten obtained her A licence than, like Amy Johnson, she started looking for a sponsor for a flight to Australia. Having failed to arouse any interest in England, she went back to New Zealand, where there was equal disinclination on the part of friends, relatives, and everyone else she approached: as she had done only a few hours' solo flying, understandable doubt was expressed about such a long flight alone. She returned to England to work for her commercial licence and the prestige which would come with it. Although flying at Stag Lane was sponsored, she had to supplement her small allowance by pawning everything possible to pay for the hundred hours of solo flying required for the B licence.

It was a fellow club member who made Jean's first attempt to reach Australia by air possible: they were to have half-shares in a second-hand Moth, and would split the proceeds both of the flight itself and of a year's tour giving passenger flights in Australia and New Zealand. It was several months before she had completed her preparations. In April 1933 she flew non-stop to Rome, nearly 1000 miles, but after a major engine failure in India the attempt had to be abandoned, leaving her in debt and with no way of raising the money to carry on. Lord Wakefield, whose generous interest in aviation had already helped many aspiring record-breakers, including Amy Johnson, came to the rescue, first by arranging for Jean to travel back to England, and then with the finance for a second attempt.

A year later, she left the man to whom she had just become engaged and took off again in a £260 five-year-old Gipsy Moth, but this time she reached only Rome. Between Marseilles and the Italian capital, she battled against strong headwinds and, just short of her destination, ran out of petrol, making a night landing on a field surrounded by wireless masts. In the daylight she realised how lucky she had been to put down safely, only 25 yds from the high embankment of the river Tiber, after somehow missing both the 100-ft masts, and the high-tension wires which bordered the field. A week in Rome ruined the chances of establishing a women's record to Australia, so she turned round, flew back to England, and set out again two days later.

It was third time lucky. In the air, Jean's time was fully occupied steering a compass course, checking her position on the map, making up the log, pumping petrol by hand, and trying to have an occasional cup of coffee and a sandwich. Although she always carried milk tablets, raisins, barley sugar, and concentrated meat tablets, she had a healthy interest in food and liked to eat regularly rather than rely on emergency flying rations. Across the English Channel and over northern France, her hands on the control column in the draughty open cockpit were numb in spite of her leather helmet, goggles, heavy lined flying-suit and fur gloves. She flew into a gale which halved her normal speed of 80 mph as she approached Athens, and over India noticed an oil leak which made her wonder if she would reach Calcutta. There was nowhere to make a landing in the dense trees on the hills beneath her, and the side of the Moth was covered with oil, with less than two pints left in the sump, by the time she arrived.

As she flew from Rangoon towards Victoria Point, the southernmost tip of Burma, she met first several severe squalls, and then early monsoon rain which completely blotted out sight of the land. The open cockpit was soon almost flooded, and her tropical flying suit was wet through. The engine protested with occasional splutters, and when, nine hours after leaving Rangoon, she saw the aerodrome through a brief break in the rain, it looked more like a lake. As she landed, throwing up great sprays of water, she was directed to the 'dry patch', where the water was only a few inches deep. In Singapore, her Dutch host for the night sliced a path through thick fog by driving several times up and down the runway, so that she took off between two impenetrable fog barriers.

The Timor Sea, on the last leg to the Australian coast, seemed endless, but when she landed at Darwin she had beaten Amy Johnson's time by four days, although she had taken 22 hours longer than her ambitious target of 14 days. Over the daunting vastness of the Australian interior to Sydney she gratefully accepted an escort aircraft, which carried her accumulating congratulatory telegrams in a borrowed mail bag. Rapturous crowds greeted her every time she landed to refuel or spend a night, and sixteen aircraft, including the famous *Southern Cross*, met her over Sydney Harbour Bridge. That evening, she broadcast to

Australian, British and New Zealand radio stations.

Unlike Amy Johnson, Jean Batten enjoyed the overpowering hospitality of Australia and the even greater excitement of the crowds in New Zealand, where, after crossing the Tasman Sea by ship – it was too far for the tiny low-powered Moth – she made a six-week aerial tour, giving speeches and being entertained by civic authorities, aero clubs, women's sports' associations, schools, hospitals, and as guest of honour at a Maori feast at her birthplace, Rotorua.

A few months after her flight, the England to Australia Air Race set out from Mildenhall for Melbourne. Jean Batten's new status as an international pilot was endorsed when she was asked by the Gaumont British Film Company to be Melbourne's radio commentator. She knew all the control points on the route: Baghdad, Karachi, Allahabad and Singapore. For the ten days of the race she broadcast several times a day, taking in the contents of cables and telegrams which came in while she was on the air, and filling in the time when there was little new information by talking about the route, the problems and dangers likely to be encountered, and the weather. Out of sixty-four entries, twenty started, two were killed and eleven, all men, finished. The winning aircraft was a Comet, piloted by Charles Scott and Tom Campbell Black, who had taken a few minutes under three days: Jean clung to her microphone and continued to broadcast as the crowd surged forward around her to greet them.

The sponsor of the winning Comet offered Jean 'some little memento, to remember the race in years to come'. She liked to be smart in a white flying suit even in the air and to change into something feminine when she landed, so chose a blue muslin de soie French model evening gown, which she christened at a ball in Sydney when the Duke of Gloucester inaugurated the first air mail service between Australia and England a few weeks later.

Flying dominated her life, and she wrote to her fiancé in England explaining that, as she did not feel it could be combined with matrimony, she wished to break off their engagement. Somehow the press heard of the broken romance and reporters turned up to interview the unfortunate man before the letter arrived. Jean had decided to fly back to England in her elderly Moth in time for King George V's jubilee, and met the hurtful side of publicity just before she set out for the return flight when she found her private life a matter of speculation in the press: 'Will Jean Batten marry?'

Although the Moth had been thoroughly overhauled, the engine gave continual trouble on the way back. Over the Timor Sea, it faltered, coughed, and died. As she glided helplessly down from 6000 ft, emerging from cloud above a featureless infinity of blue sea at 3000 ft, she undid her shoes and flying suit and reached for the hatchet she carried in case of emergency: her only hope of survival was to cut away a wing and float on it. At what seemed the last moment, the engine burst into comfortingly noisy life again.

On the final leg, across France, the cold in the open cockpit was so intense that she made an unscheduled landing to spend her last two francs on a cup of coffee. She reached Croydon 17 days 15 hours after leaving Australia, the first woman to make the flight in both directions. The next morning she made her first broadcast from the BBC. To the disgust of a minority, which included the editor of the *Aeroplane* magazine, Jean was given a contract with Gaumont-British for three talks a day for two weeks to accompany a recruitment film for the RAF: perhaps an odd choice of publicist for an air force that was determinedly all-male. She gained at least one recruit, however: the operator at the cinema.

Jean Batten had earned enough in Australia and New Zealand by lecturing, broadcasting, writing, giving passenger flights and advertising, to upgrade her Moth. For the long-distance records she was planning to break, she needed a faster, more sophisticated aircraft than the open Moth, with its limited range, 100 hp engine, exhausting manual petrol pump, and heavy wooden propeller. Her aim was to fly direct and fast enough from England to New Zealand to make the idea of a regular air link attractive. On the way, she intended to break the record set by the Australian pilot Jimmy Broadbent, who had flown to Australia from England in under a week in a Percival Gull.

On her twenty-sixth birthday, she took delivery of a new Gull, joining those who put their faith in the future of the monoplane rather than the traditional biplane. As well as an automatic petrol pump and a metal propeller, the Gull gave her the luxury of hydraulic brakes – the Moth had no brakes at all – landing flaps, and a 200-hp engine. By the time she had had extra petrol tanks fitted instead of the usual passenger seats, the total cost was

£1725: almost her entire savings as well as the proceeds from selling the Moth at a slight profit.

She said nothing about her long-term plans, but calmly spent the next few months preparing for a crossing of the South Atlantic: only Jim Mollison had flown solo from England via West Africa and crossed to South America. Her programme of physical training, which included skipping and walking, ensured that she was far fitter for the ordeal than the heavy-drinking, fast-living Mollison, who saw no need for abstinence from alcohol before or even during a flight. She had to comply with the regulations of eight countries on the 8000-mile route. The French insisted on a revolver for self-protection over the Sahara – but the South American countries would not let her in if she had any firearms with her. Recent regulations about international flying included a long list of conditions and compulsory equipment for flights to French West Africa: mooring equipment, a signal pistol with red and green cartridges, glass tubes containing chemicals which on impact with the ground would show wind direction, and a 20 per cent petrol margin. In case she had to land in the wastes of the Western Sahara, she had to find room for several large water containers and a fortnight's emergency rations, and take out an insurance policy for 100,000 francs towards the cost of a search.

She took off from Lympne on Armistice day, 11 November 1935, and landed 9¾ hours later at Casablanca, nearly 2000 miles away: an unintentional non-stop record. There was a bureaucratic delay in Casablanca because she had not protected herself against *peste*: she eventually signed a paper declaring that she was carrying on at her own risk. As she came in to land at Thies in Senegal next day, 1000 miles further on, the Gull's throttle lever jammed: two attempts to cut her speed failed, and on her third approach she switched the engine off and glided down apprehensively. She landed, following the advice she had been given, at Thies rather than Dakar, but discovered that Dakar's runways were after all complete, while Thies was neglected and overgrown, and that her petrol supplies had been sent to Dakar. With an unusual flash of anger, she insisted that something must be done immediately: at 10.30 that evening, the fuel arrived by lorry as she and the local mechanics were finishing the daily engine schedule.

Instead of resting before attempting the South Atlantic record,

Jean decided to aim for the fastest time over the whole route. To lighten the aircraft she removed her engine spares and tool kit, firearms, cartridges and drums of water, but carefully folded her two evening dresses and put them back into the aircraft. She took off in drizzle well before dawn: loneliness almost overwhelmed her as she left the African coast in the dark for Brazil, nearly 2000 miles away across the ocean.

Over the Equator, she experienced the freak effect of an electrical storm in the doldrums: suddenly the compass needle swung half-way round the dial, although the bank-and-turn indicator assured her that the aircraft was still on a steady course. The idea of struggling for another 1000 miles with no compass, and with no landmarks to guide her even had the visibility been better, sent perspiration trickling down into her eyes: her body was taut with strain as the needle swung back to its original position, the darkness outside the cabin gave way at last to light and the rain stopped thundering against the windows. Eleven hours out from Thies, she saw a ship, a welcome sign that she was on the correct course, although she longed for radio so that she could confirm her position.

After 13 hours and almost 2000 miles of ocean, she made landfall within half a mile of the point she had been aiming for, and was greeted enthusiastically in Port Natal in Brazil fifteen minutes later. She had beaten Jim Mollison's record from England by almost a day, and heard the announcement of her achievement over the radio from London a few hours later after bathing and changing her white flying suit for one of the dresses she had insisted on taking with her.

For the next three weeks Jean Batten toured South America in her Gull, delighted by what she saw and delighting the people she met. By the time she left by ship for England, she had met presidents, been the guest of innumerable dignitaries, and amassed medals, honours and gifts. She left Rio de Janeiro the owner of a diamond set in a platinum brooch from the British community, an aquamarine from the Brazilian girls at the Department of Aviation, the gold and blue Order of the Southern Cross never previously given to a British woman not of royal birth, and a bronze statue of a woman poised on a globe holding an olive branch and a scroll with the words *Conquête de l'Air*. In Buenos Aires, where she was greeted by a vast crowd which burst through the barrier of mounted police to carry her shoulder high, she was given bouquets of flowers by one of Argentina's leading airwomen, a medal to commemorate the first flight from England by a woman, and a complete wardrobe of new clothes. She broadcast in both English and Spanish in Uruguay, and was made an honorary officer of the Air Force in all three countries.

In February 1936, as the guest of the Aero-Club de France in Paris, Jean met many of the French leaders of aviation. At a dinner party given by the now elderly Blériots, Madame Blériot described how she had watched her husband's historic Channel crossing in 1909 from the deck of a battleship, showed her his workshop and explained that the walls and ceiling of his bedroom were painted blue so that he could wake every day to blue skies. Aviation had progressed rapidly since the first aerial sea crossing, and at the other end of the scale she met the record-breaking French pilot Marie-Louise Bastié. Mademoiselle Bastié had set a world endurance record in 1930 by circling Le Bourget airfield for 38 hours, made a non-stop flight in a light aircraft of 1800 miles across France, Germany and Russia in 1931, and in 1934 was the first woman to fly to Tokyo from Paris and back. A few months after she and Jean Batten met, she broke the New Zealander's record across the South Atlantic by one hour.

Whatever Jean Batten did attracted attention, and when she took her mother on a month's holiday in Spain, the idea of mother and daughter flying together caught the popular imagination. The leisurely holiday was briefly interrupted when Jean flew to Paris to receive the *Légion d'honneur*. She felt that her 'cup of happiness was indeed filled to the brim', and was ready for her ultimate test, the first solo flight to New Zealand.

In preparation she went with her mother on an 80-mile walking tour along the South Downs, a variation on her usual tedious training of skipping and daily walks. The Gull's 80-gallon auxiliary tank, which had been temporarily replaced by a comfortable passenger seat, was put back, and she treated it to a self-starter to save the time spent swinging the propeller by hand. Before she left, yet another medal was added to her proud display, that of Commander of the British Empire.

Jean Batten set out on 5 October 1936 to fly from one extreme of the Empire to the other. Although it was 3.30 a.m. when she arrived at the aerodrome, a crowd of reporters and photographers had gathered to see her off, and just before she left she impatiently made a speech in the blinding lights of a film crew. The engine was already warming up when the cameraman begged her to repeat her speech: he had forgotten to switch the sound on. Over the Channel, she suddenly realised the enormity of the task she had set herself: a flight of 14,000 miles, including the last section of over 1300 miles across the Tasman Sea, which had an unmatched reputation for treachery and danger.

In less than six days she was in Australia: she had flown through heavy rain and the dust and sand clouds of desert storms; the Gull had been battered by a typhoon and Jean was soaked as water streamed into the closed cockpit through a small leak; she had twice flown at night, and had never had more than a few hours' sleep. Before her final take-off, an over-enthusiastic band of helpers had pushed the Gull backwards, puncturing a tyre, which had been filled with all the rubber sponges from the neighbouring village to allow her to carry on.

She had beaten Jimmy Broadbent's record by more than a day, and read the many telegrams and cables of congratulation which were waiting for her in Darwin as she flew on across the lonely emptiness of Australia. In Sydney, which she reached in the record time of eight days from England, her reception was tumultuous, but tempting though it was to stay and accept the invitations which poured in, including one offering several thousand pounds for a tour of the country, she was determined to carry on across the Tasman Sea to New Zealand. She spent two and a half days in Sydney, resting after her flight of nearly 13,000 miles while her aircraft was thoroughly checked at Mascot, and trying to obtain the permission of the Civil Aviation Authority (CAA) for the last stage of her attempt to link England and New Zealand by air. The CAA refused at first to allow her to take off as the aircraft, with nearly half a ton of petrol, would be considerably overloaded, but relented when she produced a specially endorsed Certificate of Airworthiness obtained with foresight before leaving London.

It was not only the CAA which had doubts about the flight to New Zealand. She was put under considerable pressure both by friends and in the press not to carry on. The day before she left she was given a lifejacket, and was criticised in a Sydney newspaper because of the expense in time and money if she came down over the sea. Just before she took off, she made a brief statement: 'If I go down in the sea no one must fly out to look for me. I have chosen to make this flight, and I am confident I can make it, but I have no wish to imperil the lives of others or cause trouble or expense to my country.'

Her wish would no doubt have been ignored, but she reached Auckland 10½ hours after leaving Sydney: her welcome completed the triumph of the flight. It seemed as if the whole of Auckland must be there to greet her, with a 13 mile traffic jam from the airfield to the city centre. New Zealand's heroine was plunged into an exhausting tour, for which she felt she must find the strength both to meet popular demand and to clear her expenses so that she could afford to continue flying. She had however pushed herself almost to the limits of endurance in the 11½ days it had taken her to fly from England to New Zealand, and after a few days reluctantly cancelled the rest of the tour.

While she rested as the guest of the New Zealand government, a fund of £2000 was collected for her. When she visited Rotorua with her parents, she was again guest of honour at a Maori reception, where she was given a chief's kiwi feather cloak and the name *Hine-o-te-Rangi*, Daughter of the Sky. A few months later, after her return by ship to Australia, she joined the search for a Stinson airliner which had disappeared on a flight from Brisbane to Sydney with nine people on board, one of whom was her third and last fiancé.

When, after several leisurely months in Australia, Jean decided to return to England, the obvious way to do so was in her own Percival Gull. Between dawn on 19 October and 24 October 1937, she cut Jimmy Broadbent's recent record of 6 days 9 hours by more than 14 hours, and had the distinction of holding the world record between England and Australia in both directions. Banquets, luncheons, receptions, invitations to clubs and societies followed one after another. She broadcast on radio and, for the first time, on television, met Colonel Moore-Brabazon, the first Englishman to gain a pilot's licence, Gracie Fields, the pioneering pilot Sir Alan Cobham, numerous lords and ladies, Air Vice-Marshal Baldwin, King Leopold of Belgium, King George VI and Queen Elizabeth, and 'little Princess Elizabeth'. In January 1938, she was given the world's highest aviation award: the medal of the *Fédération Aéronautique Internationale*. Twenty-two nations had participated in the voting: she was the twelfth aviator and the first woman to receive the medal. She had already won the prestigious Harmon Trophy three years running, once sharing it with Amelia Earhart.

Jean Batten's fame was international. The dentist's daughter from Auckland had, without ever losing her popularity or becoming conceited, experienced time after time the joy of achievement which for her was 'the greatest and most lasting of joys'.

At the start of the Second World War she gave up flying, although she raised money for war funds through lecture tours, and afterwards moved with her mother first to Jamaica and then to Spain. On her seventieth birthday she celebrated by doing the can-can, kicking higher than anyone else, and was last heard of in August 1982, when she was seventy-two and was about to leave Malaga for Majorca. All attempts by the Foreign Office, the New Zealand High Commission, Prime Minister Robert Muldoon in New Zealand, and the publisher of her autobiography, failed to trace her. Her passport expired and her mail piled up. The Daughter of the Sky was officially 'a missing person'.

10

Nancy Bird:
Australian Mercy Pilot

From the age of thirteen, Nancy Bird's sole ambition was to fly, and her favourite book was a flying manual. For three years, she saved for flying lessons from her meagre earnings as house-keeper to her father, and bought herself a leather flying-jacket, goggles ordered from a catalogue, and a helmet specially made to fit her small head. She was seventeen, weighed only a little over seven stone, and could reach 5ft 2in. only by standing very straight, when she had her first lesson: to see out of the Cirrus Moth she had to sit on several cushions, which later were the bane of her flying life.

In August 1933 Charles Kingsford-Smith started a flying school at Mascot, later to become Sydney's international airport, but then no more than a flat area, 800 × 500 yds, with one red dirt runway on 160 acres of reclaimed swamp. Nancy – an ardent admirer of the pilot who by then had made record flights to, from and round Australia, as well as round the world – became one of his first pupils at Mascot. She walked away after her 20-minute inauguration beaming because he had told her: 'All right, not bad for a first time. You'll learn.'

While she was learning to fly, Nancy gained a new awareness of the beauty of earth and sky, and a practical knowledge, considered somewhat eccentric for a woman, of aircraft mainte-nance, radio and semaphore. To reach the airfield every morn-ing, she had to cross Sydney Harbour by ferry, arousing attention by her extraordinary clothes: as she had soon discovered, the fussy dresses more usual in Australia in 1933 were not suitable for open cockpit aircraft, and she wore knickerbockers of linen for the summer and of tweed for winter, with long socks and flat shoes, and did her hair in a bun on the nape of her neck. Pilots were exotic, heroic creatures, but a woman pilot was a distinct oddity: even Kingsford-Smith did not really approve of Nancy's wish to fly.

As soon as she had her A licence, she started working for her commercial licence with seven other pilots, all male except one, Peggy McKillop, two years her senior. She flew every aircraft she could: a strangely unbalanced rebuilt Avian, an old Curtiss Jenny, the Avian Red Rose in which Captain Lancaster and Mrs Miller had arrived from England in 1928, Puss Moths and Gipsy Moths. Nancy and Peggy were the first women in Australia to fly at night, and when Nancy was issued with Commercial Licence Number 494 in March 1935 she was the youngest woman in the British Empire to be qualified to fly commercially. With £200 from an aunt and £400 from her father, she bought and rebuilt an old Moth which a certain Lady Chaytor had flown out from England in 1932.

In April 1935, the two girls, who were soon known as the Little Bird and the Big Bird, embarked on the first barnstorming tour made in Australia by women. They had worked out an itinerary with the help of the Shell Oil Company, which was to send press releases ahead of them and arrange fuel supplies, but the male pilots were depressingly sceptical about the venture. Barnstorming had been played out, they warned: even the famous *Southern Cross*, the trimotor Fokker in which Charles Kingsford-Smith had crossed the Pacific and the Tasman Sea, had ceased to be a money-spinner at air pageants, although in the early days he had been able to charge joy-riders £5 a flight. The girls would, moreover, not be able to offer the spectacular entertainment which had been provided by some of the men. Charles Kingsford-Smith and Harold Durant, both ex-war pilots, had staged mock air battles between two aircraft bearing respectively Royal Air Force markings and the Iron Cross of Germany. Kingsford-Smith's nephew John had given displays of wing-walking and aquaplaning behind a seaplane, and Jim Mollison and Hereward de Havilland had thrilled the crowds by diving at each other and swerving apart only just in time to avoid a mid-air collision.

Before leaving Sydney, Nancy was £30 in debt until the editor of a new magazine, *Woman*, offered £10 a week for three weeks if the aircraft, christened *Vincere* – to conquer, in Latin – carried the word WOMAN painted under its wings. The idea of two girls

2 Dolly Shepherd, Edwardian 'Queen of the Air' in her parachuting costume

Mrs Letitia Ann Sage, accompanied in Lunardi's balloon by George Biggin: the first British woman to make an aerial ascent in 1785

Dolly Shepherd, astride the sling and gripping the trapeze bar, before an ascent under a balloon, Wolverhampton, 1910, while Capt Gaudron settles passengers in the basket; the parachute for Dolly's descent from 2000 ft is laid out on the ground in readiness

4 Gertrude and John Bacon, after a rough landing on Welsh mountainous coastline following a nocturnal ballooning expedition in 1899; Gertrude broke an arm and her father's clothing was badly ripped

200.000ᶠ de PRIX

GRAND
SEMA

S Montaut

D'AVIATION

5 At the end of the Grande Semaine d'Aviation, the first flying meeting, held in Reims in 1909, Gertrude Bacon became the first British woman to be flown in a powered aircraft

6 Harriet Quimby (*r*) and Matilde Moisant, the first women in America to gain pilots' licences, soon after learning to fly in 1911; in the air Harriet Quimby wore a purple silk flying-suit, and Matilde Moisant a heavy tweed knickerbocker outfit with knee-length boots

7 Gustav Hamel gives last-minute advice to Harriet Quimby before the first solo flight by a woman across the English Channel on 16 April 1912; her Blériot monoplane had no brakes, so six men had to hold it while the 50-hp Gnôme engine was running

8 The Stinson sisters, Katherine and Marjorie, in a Wright biplane, 1913.
Katherine's aerobatic flying financed the Stinsons' family flying school, and
Marjorie trained Canadian pilots during the First World War

9 Mabel Cody's Flying Circus specialised in high-speed stunts which
included transferring from a speedboat to an aircraft and required split-
second timing, 1927

10 Lady Heath stepping out of her Avro Avian at Croydon, wearing the clothes in which she landed after her flight from Cape Town in 1928

1 Lady Bailey, who, in 1928, flew solo in a de Havilland Moth from London to Cape Town and back

12 The Hon. Mrs Victor Bruce, with her Jowett car and the Fairey Fox in which she participated in a popular flying circus known as the British Hospitals' Air Pageant

13 The Duchess of Bedford with Captain Barnard, pilot, and R. Little, navigator, in front of the Fokker in which they flew to India in 1930

14 Ruth Elder, actress and pilot: her attempt to fly the Atlantic in 1927 finished in the sea, but gained her considerable publicity

15 Line-up of aircraft at Douglas, Arizona, during the first Women's Air Derby held in 1929; Amelia Earhart's Lockheed Vega in foreground

16 Amelia Earhart, first woman to cross the Atlantic by air, with Bill Stultz,
    pilot (*r*), and Louis Gordon, co-pilot, at their official welcome to Burry
    Port, Wales, 1928

17 Amelia Earhart looking out from the *Friendship*, in which she was flown
    across the Atlantic in 1928; on tow on Southampton Water after the flight

18 Amy Johnson, whose solo flight
   from England to Australia in 1930
   made her a national heroine

19 Nancy Bird, one of the
   inaugurators of the Australian
   Medical Air Service

20 Jean Batten, the New Zealander
   who made the first solo flight in 11½
   days [in 1936] from England to New
   Zealand

21 Pauline Gower, Commandant
   of No 5 ATA Ferry Pool at
   Hatfield, Herts, during the
   Second World War

22 Charles and Anne Lindbergh in France in 1933; her role as the wife of the famous pilot included learning to fly

23 Beryl Markham, the Kenyan racehorse trainer who, in 1936, made the first flight across the Atlantic from England to America, in a Percival Gull, with (*l*) Edgar Percival

24 Air Transport Auxiliary women pilots, including Pauline Gower (*l*), and de Havilland Tiger Moths at Hatfield, 1940

25 ATA pilots enjoying a break at Hatfield, 1940; behind them is a de Havilland Flamingo

26 Hanna Reitsch, German test pilot, wearing the Iron Cross, Second Class, and reading an account of its presentation to her by Adolf Hitler at the ceremony in 1941

27 Jacqueline Cochran, who learnt to fly to further her business interests and broke innumerable flying records; the first woman to break the sound barrier in 1953 while flying a Canadian Sabre jet

8 Jacqueline Auriol with test pilot Carl Patton in front of a Lockheed T-33 jet trainer during a tour of US aircraft factories and military installations

29 Sheila Scott, the first woman to fly solo round the world by the longest possible route (Equator) as well as over the North Pole

30 Judith Chisholm, commercial pilot, who made a record-breaking solo round-the-world flight in 1980

31 Lynn Ripplemeyer, first woman captain of a Boeing 747 jet, 1984

32 Air UK all-female crew, 1984 (*l–r*): Mandy Ackroyd, First Officer; Pat Richardson, Captain; Stewardess No 2 (temporary); Claire Wilson, Stewardess No 1

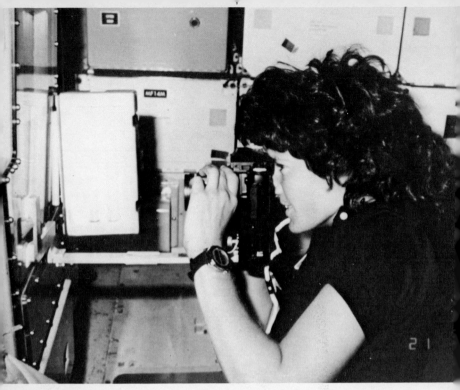

33 Dr Sally Ride, first American woman astronaut to venture beyond the earth's atmosphere, in 1978; in 1984 she was a mission specialist member of the crew of the earth-orbiting *Challenger*

34 Dr Kathryn Sullivan, mission specialist, floats on board *Challenger* to join fellow astronauts; in 1984 Dr Sullivan was the first woman to walk in space

touring with an aircraft proved new enough to be glamorous, although some of their passengers were very nervous: the little Moth looked even smaller and more fragile in the vast open spaces of Australia, where the children were often given the day off school to see what was sometimes the first aircraft ever to land near their town. Peggy and Nancy stayed in 'two-bob hotels' where a bed, usually in a shared room, breakfast, lunch and dinner, each cost two shillings. Because of the restricted space in the Gipsy's luggage compartment, Nancy's wardrobe for three months consisted of two changes of underwear, two pairs of stockings, two blouses, a flying suit and leather coat, a long uncrushable skirt and a beaded lace overblouse: in the skirt and blouse, she went to grand picnic race-meeting balls which were the highlight of the rural social calendar.

Although it was a tough life, flying from place to place over the country to the north-west of Sydney, they managed to make enough money to eat and to maintain their aircraft. They flew 22,000 miles during their three-month tour, although *Vincere* was often temperamental about starting. In the autumn, Nancy accepted an offer from a wealthy landowner in New South Wales: Mr Perry, who had asked her to fly his son Dick 50 miles to look at some sheep, suggested she should borrow the money for a larger and more reliable aircraft suitable for charter work. He owned land in New South Wales both at Narromine and at Dubbo, where he hoped to provide Southern Air Lines with a landing field. Between her first and second barnstorming tours, Nancy was given the task of selling shares in the company and attempting to persuade Dubbo that it needed a regular air service: the mayor was convinced that it could never compete with the overnight train service to Sydney.

It was at Dubbo that Nancy met the Rev. Stanley Drummond, who had set up a Far West Children's Health Scheme: many of the children of that part of New South Wales suffered from trachoma, a disease brought on by glare, dust and malnutrition, and which led to blindness. He had managed to bring some of the worst cases to Sydney for hospital treatment, had arranged holidays for some who had never left their isolated and poverty-stricken homes, and had bought obsolete railway carriages for conversion into mobile clinics. Many families lived hundreds of miles from the nearest railway. It took several days for nursing sisters to reach the remote areas, travelling by car with nowhere to stay on the way, in temperatures as high as 116°F in the summer, or unable to carry on along boggy tracks and roads in wet weather. Baby health clinics had been established, and as an experiment Stanley Drummond asked Nancy to fly the sister based at Boruke in New South Wales on her round of visits to the tiny settlements many miles to the west of the rail terminus.

The second barnstorming tour was reorganised to find time both for the experiment, and for Narromine's 1935 air pageant, at which Peggy and Nancy entered a closed-circuit aerial derby run on Caucus race lines. As all entrants broke their time allowances, the race was awarded to those who did so the least, which, as the times were the pilots' own estimates, meant that those who had told the smallest lies won: the girls were awarded a silver cup. From Narromine, they flew over a sea of timber towards Bourke in northern New South Wales: for 240 miles there were no visual landmarks, nowhere to make an emergency landing, and the sheep, scattered one to over 12 acres among the trees, looked from the air like grains of rice. Bourke seemed like the back of beyond to a city dweller: to those who lived hundreds of miles beyond it to the west, it was a city.

Two days after her twentieth birthday, Nancy Bird took off with Sister Webb on her first Far West Children's Health Scheme tour. They flew over country which had had no rain for over eighteen months, stopping at places where it was impossible to get fresh fruit or vegetables, where the nearest telephone was fifty miles away and the nearest hospital 100 miles further, and where house-proud women fought a constant battle with blowing dust which often also obscured the roads. By the time the second tour, and the first 'Mercy Mission', were completed, Nancy had made the final arrangements by letter for her new aircraft. She was offered a contract with the Far West Children's Health Scheme, with a six months' retainer fee of £100 and a guarantee of £100 worth of work. In return, she would base her aircraft at Bourke and make it available for clinic tours, but would otherwise be able to take on charter work to earn much-needed extra money towards the loan for the £1700 aircraft and its upkeep. She accepted: it was both a way to help the women who were rearing families in conditions of considerable hardship in the outback, and a job in aviation.

After the open cockpit of the Gipsy Moth, the new Leopard Moth's cabin with triplex windows, heating and ventilation, and two passenger seats behind the pilot's bucket seat, all upholstered in grey leather, seemed luxurious. Storage pockets and a generous luggage locker were another welcome luxury, as was the cruising speed of 120 mph instead of the Moth's 80 mph. Nancy was so light that ballast had to be added, and had flown the Leopard for only two hours when she left Sydney alone to fly over the Blue Mountains at 6000 ft and then, in a thick dust haze, over the vast expanse of timber forest to Bourke. The excitement of setting out with her new aircraft to start her first commissioned flying job was marred by the death of Sir Charles Kingsford-Smith, who disappeared on a flight from England with a co-pilot.

The first concern when she arrived was to arrange some shelter for the aircraft: scorching sun could damage the paintwork and shrink the wooden frame, and sudden freak winds, which were common, could wreck an aircraft even when it was tied down. The next task was to find somewhere to live: when the owner of a hotel discovered who she was and why she was in Bourke, she offered her a room and full board for only £2 a week. It was unusual for a girl to live alone in a hotel, and gave rise to some initial comments about her respectability, not helped by her habit of wearing shorts and a topee.

The only maps available were those given away by garages: accurate enough for a motorist, but not very helpful to a pilot flying at 120 mph. Roads and railway lines, the navigational aids of European pilots, were few and far between, and the Overland Telegraph Line, a single wire on a line of iron posts, was the only landmark easily identifiable from the air. It could even be literally a lifeline: not only could a pilot look for it when lost, but cutting the wire after an emergency landing guaranteed that within a day or so a linesman would come out to mend the break. Courage, determination and stamina were needed to fly over the Australian outback, with no radio, a single engine, and hundreds of waterless miles between habitations. There was no one to check up if a pilot failed to arrive at a destination, which could be unknown to anyone else, and little chance of survival if an aircraft came down. It might be days after a forced landing before a search party was sent out, and even longer before it found the aircraft and pilot. In 1929, two pilots, Hitchcock and Anderson, had died of thirst after drinking even the spirit from their compass. Nancy always carried a large unbreakable vacuum flask of water, as well as emergency rations of raisins, barley sugar and chocolate.

Her clinic flying was interspersed with charter work, in which she had to trust to the descriptions of her charterers about suitable landing grounds, often barely cleared scrubland with holes and stumps. In good weather, charter jobs, at a shilling a mile, were often few and far between, and she would fill in time by taking potential customers for joy-rides; in bad weather, when the roads were swamped and the telephone lines down, she had to fly long hours to fit in all the charter trips, and could often have employed three pilots and three aircraft. The temperature in the Leopard's cabin, after it had been standing outside, often reached well over 100°F, and flies were a constant problem. Rain, which brought the bare countryside to life and made flying over it afterwards a pleasure, brought mosquitoes and the problem of landing on waterlogged ground or rescuing stranded motorists. Many people in Bourke were under the misapprehension that Nancy was a wealthy young woman who flew for fun, an impression she did not try to dispel: she hated to admit that she was so hard up that she wore her sister's cast-off clothing and relied on vitamins sent by her father to supplement her meagre diet.

In December 1936, Nancy competed in the Brisbane to Adelaide Air Race, enjoying the company after the months of being a lone pilot. She was one of five women to enter, and had to contend with fog, rain, strong headwinds and turbulence over scrubby sandy desert to win the Ladies' Trophy by one minute. The outright winner of the race was a garage proprietor from Hamilton, Victoria, Reg Ansett, who later became managing director of Ansett-ANA, one of Australia's largest airlines. Nancy stayed in Adelaide for a fortnight to take part in an air pageant and derby, and then moved to Queensland: the Far West Children's Health Scheme could no longer afford to pay her retainer, and without the clinic flying there was not enough charter work in Bourke to give her a living.

She failed to persuade the Queensland government to use her for aerial ambulance work, although the Australian Inland Mission based at Cloncurry and the Bush Church Aid Society in

South Australia were both by then operating aerial ambulance services. The costs of living at Charleville were greater than they had been at Bourke, and when the president of the Cunnamulla branch of the Red Cross, who also happened to own the Cunnamulla Hotel, offered her free accommodation, Nancy moved again. She continued to do charter work, including private ambulance flying, taking doctors to patients and patients to hospitals.

Operating alone, without insurance, she was constantly worried about the safety of her passengers and struggling to meet the costs of her aircraft. The strain was affecting her health, and after a series of infected throats, and an operation to remove her grumbling appendix before it erupted – possibly in some remote place where she would have been the only person capable of flying herself to hospital – the day came when Nancy could not face flying through the severe turbulence caused by the extreme ground heat. She put her Leopard and its hangar up for sale. By the time they had been sold and her debts paid off, she had £400 left, approximately the amount she had started with. In the meantime, however, she had to her surprise and gratification become famous for her intrepid mercy flights.

As the guest of Dutch, German, French, Danish and Swedish airlines, Nancy embarked on a world tour, travelling 45,000 miles and visiting twenty-five countries, alternating between luxurious hospitality and penny-pinching economy: in England, she was presented at Court one day, and the next had baked beans for fourpence, the cheapest thing on the Lyons Corner House menu. In America, she flew on the maiden flight of the DC4 with Jacqueline Cochran, and when she returned to Sydney she mounted a successful historical exhibition, 'Wings of the World'. She married in 1939, and during the Second World War turned down requests for her services with the Women's Australian Auxiliary Air Force and the Air Transport Auxiliary in Britain because of her family commitments. With the Women's Air Training Corps she nevertheless compiled a list of Australian women pilots who might be used as ferry pilots in Britain: by the time the register was complete, the war was over.

In 1949, Nancy Bird Walton organised a gathering of Australian women pilots, from which grew the Australian Women Pilots' Association of which she was founder-president. Twelve years later it had under 200 members: it was, she felt, only in America that women had the economic independence to fly for fun. Twenty years after selling her Leopard and giving up flying, she renewed her private pilot's licence in a Tiger Moth, in which she was at first terrified: she had been invited to participate with a co-pilot in America's annual women's cross-country race. In a hired Cessna 172, which she had first to learn to fly, as well as learning how to operate the radio, she came fifth, and entered the Powder Puff Derby again in 1961, in a 250-hp Piper Comanche: it flew at twice the speed she had been used to and had over twice the horsepower.

In 1985 fifty women described by the press as 'birdladies' attended a three-day seminar of the International Women's Airline Pilots Association, an organisation with 120 members, nearly threequarters of the world's female commercial pilots. The seminar was sponsored by Ansett, which employed five women pilots but which only nine years previously had fought and lost a battle in the high court to exclude women from its pilot training course. Nancy Bird Walton, the first woman in Australia to operate an aircraft commercially, attended the seminar's press conference: she 'beamed as she heard how far down the professional track today's women pilots had gone'.

≪ 11 ≫

Beryl Markham:
Colonial Huntress

In 1936 a Percival Gull landed in a bog in Nova Scotia. Its pilot, Mrs Beryl Markham of Nairobi, had flown non-stop from Abingdon on the first solo flight across the Atlantic from England to America. Although she had been a freelance pilot in Africa for six years, and had flown several times between Nairobi and England, it was her first record attempt. It was also her last, and only one episode in a colourfully eccentric life: for even in Kenya, where the antics of the ex-patriate upper classes were notorious, her private life was the subject of frequent speculation. In her public life she claimed to be the first woman racehorse trainer in Africa, the first resident commercially licensed woman pilot in the continent, and the first person anywhere to hunt elephant by air.

Markham was her third surname. She had lived in Africa since 1906 when at the age of four, her father, Captain Charles Clutterbuck – a divorced, Sandhurst-trained army officer and a classical scholar – took her to what was then British East Africa. Her mother and brother stayed in England, and when her mother later visited her in Africa, Beryl found her 'an awful bore'. The only woman who had any influence on her early life was Lady Delamere, whose husband used to amuse himself at parties by shooting golf balls on to the roof. Lord and Lady Delamere were the Clutterbucks' nearest white neighbours in the Highlands where the atmosphere was said to induce euphoria in white people and where, by 1915, a thousand white farmers, many of them aristocratic, owned more than 4½ m. acres. Beryl later described Lord Delamere as 'the champion of the East African settler': he felt that the opening up of new areas by 'genuine colonisation' was to the advantage of the world, but balanced his devotion to the indigenous Masai people by entertaining 200 settlers with 600 bottles of champagne at a reception in 1928, while the black population was starving during a famine.

The Uganda Railway, which ran from Mombasa to Kisumu and took five years to build at a cost to Britain of £6 m., was opened five years before Charles Clutterbuck settled at Njoro. There he created a vast farm out of the forest, bush and rock: the railway was his chief customer. He also bred and trained racehorses, including those owned by Lord Delamere, and Beryl was brought up in an atmosphere of horses and horsy talk: race meetings at the Nairobi course were as important a part of colonial social life as was Ascot to that of upper-class England.

Beryl grew up speaking Swahili, Nandi and Masai, but had little formal education, although she was sent to various schools, from which she was expelled, and had several governesses, whom she ignored. Her companions were the children of her father's African employees, who took her on their hunting expeditions, setting out early in the morning after a ritual drink of mixed blood and curdled milk from a gourd, staying out all day, and often running into danger. Her understanding of the animals of the bush and jungle was later invaluable when she was tracking animals from the air.

Grooming and riding the horses and assisting at foaling were all a natural part of her daily life. When her father was forced out of business after honouring a milling contract in a drought, and left Kenya to make a new life in Peru, Beryl was left to support herself through her expertise with horses. She was seventeen, with only her horse Pegasus and a couple of saddle-bags containing her personal possessions: but with these, she trekked to Molo to set up as a racehorse trainer. A year later, her first winner of the Nairobi St Leger guaranteed her place in the traditionally male world of horse racing.

After marrying and divorcing a Scottish rugby player, Jock Purves, Beryl became in 1927 the wife of Mansfield Markham, who had left the Foreign Office to breed horses in Kenya. Although marriage was never more than an unfortunate interruption in her life, with Markham she did enjoy a brief period of luxury, with a honeymoon in Paris and clothes from Chanel. In London, where she lived for more than a year after her second marriage, her relationship with the Duke of Gloucester and her habit of going barefoot as his palace guest aroused considerable curiosity. The Duke, whom she had previously met in Kenya, sent her cartloads of white flowers by Shetland pony, and her husband threatened to cite him in divorce proceedings: instead, although Beryl never admitted to having had an affair with him, her royal admirer apparently settled a 'sizeable sum' on her in 1929, the year her only son, Gervaise, was born.

By then, Beryl and Mansfield were separated. Beryl felt herself temperamentally unsuited to motherhood, and returned to Kenya, leaving Gervaise to be brought up in England by his paternal grandmother. In the early 1930s her base, both socially and as a pilot, was the Muthaiga Country Club outside Nairobi, headquarters of a notorious ex-patriate 'Happy Valley Set' founded in 1924 by Josslyn Hay, the twenty-second Earl of Errol. Errol's murder in 1941, and the subsequent unsuccessful search for his murderer, uncovered the sexual complexities and extravagances of a way of life which seemed unbelievable, especially when much of the rest of the world was at war.

Long before then, however, adultery, alcohol and intrigue had figured largely in the lives of some of the white settlers who gathered at the Muthaiga Club. 'Are you married, or do you live in Kenya?' was a much quoted, and greatly resented, jibe at the ex-patriate white community of colonial Kenya. Even the hard-working white farmers, whose lives were otherwise often solitary, relaxed when they came to town at the Muthaiga Club. Opened in 1913 in a pink stucco mansion, it had tennis courts, a golf course, stabling for polo ponies, two chauffeur-driven limousines and a chef from Goa; after Kenya became independent it remained, however unofficially, almost as exclusively and expensively white as ever.

Beryl Markham, who was considered to resemble Greta Garbo, was described by the writer Martha Gellhorn as 'Circe casting a spell on Ulysses so that she could go along on his journey': 'In passing, she bewitched this company of men so that they did not resent her intrusion into their macho society, but welcomed her. It was easy to entrance the whole lot, that being her nature, and she knew what she wanted: knowledge and adventure.' Tall, broad-shouldered, blonde, with blue eyes and a 'panther-like grace', Beryl attached as little importance to her passing liaisons as to her equally impermanent marriages, although her relationship with Denys Finch Hatton, one of the men with whom her name was linked, upset the love affair between him and Karen Blixen, the famous writer, whose ex-husband Baron Bror Blixen was another of Beryl's closest friends. At a dinner party given by Karen Blixen for the Prince of Wales in 1928, Beryl was described by her hostess, in spite of the threat she posed, as looking 'ravishing'.

Several years earlier a young man called Tom Campbell Black had confided in her his ambition to make a life of flying. Beryl had already earned her place in a man's world, and when Tom Black landed at night outside Nairobi her curiosity was mixed with resentment as the sound of the aircraft's engine disturbed 'a slumber of contentment'. He had flown from London in a new aircraft which was to carry mail across Africa for Wilson Airways. Beryl satisfied her need for adventure by learning to fly, and Tom, as her instructor, passed on his knowledge of flying in the dangerous terrain of Kenya by allowing her first to get into difficulties, and only when she had felt exactly what was happening took over and showed her what to do. Beryl later claimed that, as a result, she never got out of a plane without knowing what might have happened if she had taken some other course of action.

By the time she was twenty-eight, Beryl had obtained her commercial licence and, as a professional pilot, could fly mail, medicines and occasional passengers across East Africa. Although her base was the Muthaiga Country Club, the places where she landed were often little more than hastily cleared strips in the bush. The few maps that were available for such areas showed little detail, with a scale of 32 miles to 1in. covering country which had, as often as not, never been surveyed: in Europe pilots used 4 in.-to-the-mile maps. In Kenya in daytime, the only sign of life might be a spiral of smoke, and flying on a dark night could create a feeling of unreality which made the existence of other people seem 'not even a reasonable probability'.

In 1931 Beryl Markham narrowly missed sharing the death of Denys Finch Hatton, whom she had known since she was eighteen and who, in his years in Africa, had acquired a reputation as an expert hunter to add to his various intellectual and sporting achievements. Flying was a more recent enthusiasm in which, although inexperienced, he had, according to Beryl, the same casual confidence which accompanied everything he did. Denys suggested that safari reconnoitre from the air could be developed as a service to the wealthy and often aristocratic hunting parties which had already turned Kenya into a rich man's holiday resort, and invited Beryl to go with him when he tested his theory. Tom Black, still Beryl's flying mentor, advised her against accompanying Denys in his Gipsy Moth on the first aerial elephant search. Beryl followed his advice, although it seemed illogical: the weather was not bad, and Tom admitted that the idea had possibilities. Denys and the Kikuyu boy with him in the aircraft were both killed when, for no apparent reason, it crashed after a successful trip: Beryl developed safari flying herself. By this time, she had two aircraft: a leased Leopard Moth in which she could take up two passengers for one shilling a mile each, and an Avian with room for only one passenger.

As the first person to scout elephants by air, Beryl had an ambivalent attitude. 'Flying for elephant' was a dangerous, lucrative and exciting activity, but she felt that it was absurd for a man to kill an elephant: 'It is not brutal, it is not heroic, and it is certainly not easy.' The essence of elephant hunting was, she claimed, discomfort in such lavish proportions that only the wealthy could afford it. One of those for whom she helped to make the discomfort more easily attainable was the Swedish hunter Baron von Blixen.

In March 1936 Beryl flew 'Blix' from Nairobi to London. She had already flown the route four times alone and twice accompanied, but she still considered it an epic voyage rather than a trip. With the baron she was held up several times. In Cairo they had to wait for six days for Italian permission to cross Libya. In Libya, at the border garrison of Amseat, they were again delayed until promising, for no clear reason, to circle three forts on their way to Benghazi. The third evaded them and they decided to risk imprisonment in Benghazi, the threatened penalty for not carrying out the order, rather than be lost at night over the desert. In Sardinia, they were allowed to leave only after persuading the authorities, first that Beryl was not a man in disguise, and then that they were not both spies.

In London, Beryl decided to attempt one of the few long-

distance flights not yet made by either man or woman: England to
America, across the Atlantic from east to west. She was sponsored by one of the Muthaiga set, John Carberry, an English
aristocrat who had rejected his title, adopted an American accent
and addressed Beryl as 'Burrrll'. After his education at Har-
row, Cambridge, Switzerland and Leipzig, and First World War
service in the RNAS, he had attempted to become an American
citizen, but his naturalisation papers were withdrawn because of
his bootlegging connections. Instead, he settled in Kenya as plain
John Evans Carberry. His first wife divorced him for cruelty in
1919; his second wife, Maia Anderson, was a pilot who died
when she crashed her aircraft, probably accidentally, although it
was suggested that she had sought death as the only escape
from her husband's bullying. June Carberry, Beryl's hostess at a
London dinner party, became his third wife in 1930 at the age of
seventeen, half that of her husband. She was 'common as hell'
and 'a terrifyingly unnatural blonde', with a deep voice, an
enviable capacity for alcohol, and the reputation of having a
warm heart and any number of lovers: she was later rumoured to
have been one of the Earl of Errol's many mistresses, and even to
have had a hand in his death.

At the Carberrys' London dinner party, the idea that the
Carberrys should sponsor Beryl was put forward by another
guest. They agreed to provide an aircraft for a record-making
flight provided it was something so far untried. England to
America was suggested and accepted as, when Jim Mollison flew
the Atlantic from east to west, he had shortened the flight by
starting from Ireland. John Carberry's promise of financial support was accompanied by the comment: 'Gee, I wouldn't tackle it
for a million. Think of all that black water! Think how cold it is!'

For three months, Beryl flew almost daily to Gravesend in a
hired aircraft to check on the progress of a new turquoise-and-
silver Percival Gull in which extra fuel tanks were fitted to
increase its range from 660 to 3600 miles. Tom Black suggested
that, as her financial backer lived at a farm called Place of Death,
and the aircraft was being built at Gravesend, she should call it
'The Flying Tombstone'; a facetious idea rejected in favour of *The
Messenger*.

By September, when weather and wind direction were unlikely to be favourable, Beryl was beginning to realise the dangers

and terrors of the undertaking. Even the 250,000 miles she estimated she had already flown seemed meagre preparation, but her pride would not let her back down. With Jim Mollison's watch, lent with a warning not to get it wet, a sprig of heather from a Scottish ground mechanic, and a forecast which, although not encouraging, was the best that could be expected at that time of year, Beryl took off from RAF Abingdon on 4 September 1936. Ahead of her was the prospect of flying 3600 miles non-stop, more than half of it over ocean, alone and without radio.

The Gull's petrol tanks had no gauge, and switching from one to another involved groping for a torch while for half a minute the aircraft was without power. When the first tank ran out, Beryl was flying in the dark at 130 mph towards Newfoundland into a 40 mph headwind, through a storm which had not been forecast. It was not until Beryl's altimeter told her that she was less than 300 ft above the water that the fuel from the second tank reached the engine and she was able to start climbing. By dawn, she had flown blind for 19 hours. Forty miles from land, her engine cut out, spluttered and started again. This happened several times, and the fact that visibility was at last perfect was little comfort as she struggled to clear what she thought must be an airlock.

The engine finally cut out and refused to start over Nova Scotia. Beryl landed in a bog, cutting her head as she did so, 21 hours 25 minutes after leaving Abingdon: a record flight, although because she had not managed to reach New York non-stop she felt it to be a failure. The engine was discovered to have been suffering from carburettor icing, and it seemed nothing less than a miracle that the aircraft had not come down in the sea. Beryl was rescued by a fisherman; the next day she flew into New York, in another aircraft, to a heroine's welcome. But the triumph of success was embittered by the news of the death of Tom Campbell Black, killed when another aircraft ran into his as he was taxiing ready for take-off at Liverpool aerodrome: an unnecessary death, she commented.

The flight made Beryl famous for a while, but not rich, and as she could not afford to buy the Gull from John Carberry, he shipped it back to Kenya and sold it to a wealthy Indian, who simply left it exposed to the weather on the airfield at Dar es Salaam until its engine rusted and the paint on the wings peeled. Beryl Markham never made another record attempt.

She stayed in California for several years, and married for the third time in 1942. Her last husband was Raoul Schumacher, a prolific ghost writer. He is thought to have been partly responsible for the quality of the writing in Beryl's book *West with the Night*, an impressionistic and lyrical autobiography starting with her early life in Africa and finishing with her flight across the Atlantic, but omitting incidental details such as her marriages. It was hailed as a literary masterpiece by Ernest Hemingway when it was first published in 1942: he had known her 'fairly well' in Africa, but had never suspected she could put pen to paper except to write in her log-book. His praise for her 'bloody wonderful book' was tempered by the comment in a letter that she had omitted 'some very fascinating stuff that I know about her which would destroy much of the character of the heroine'.

Beryl and Raoul Schumacher were part of a frivolous and extravagant social scene in Montecito in California. When she left him, she resumed the name of Markham in Kenya, where she became one of the country's most successful racehorse trainers, winning the top trainer's award five times and the Kenya Derby six times from her stables on Lake Navaisha. Although she was respected and accepted professionally, she nevertheless gained the reputation of being arrogant and difficult. Her success lasted until 1963, when a virus, the 'Beryl Bloom', forced her to close her stables: the horses, although still looking healthy, could no longer run. For a while she trained horses for the Countess of Kenmore in South Africa, before returning yet again to Kenya, but her career as a trainer was virtually over.

In her eighties, Beryl Markham was living in a borrowed bungalow near Nairobi racecourse, driving twice daily to feed two horses. A constant supply of vodka and cigarettes alleviated her boredom and loneliness: her glamour, helped by make-up, had survived the traumas of old age and even the onset of amnesia. She died in July 1986, only a few weeks before the fiftieth anniversary of her transatlantic flight.

12

Pauline Gower
and the Forgotten Wartime Pilots

When, on a cold February morning in 1931, a London newspaper reporter called at the Stag Lane aerodrome in search of a story, two young women who had recently gained their licences told him that they were going to fly to America together, via Baffin Land and Greenland. Pauline Gower was to be the pilot, and Dorothy Spicer would go as ballast, complete with parachute, so that if the weather was bad she could float down on to a flat-topped iceberg and wait for her partner to send out a rescue party from Canada. The story appeared, complete with its subsequent contradiction: 'The happy pair went on earnestly with their work, quite unaware that their picturesque hoax had been taken seriously in some quarters.'

Out of the hoax grew the idea of working together: Dorothy, who had completed 40 hours solo, would concentrate on the engineering, while Pauline, the daughter of Leicestershire MP Sir Robert Gower, was confident enough of her ability after 15 hours' solo flying to be chief pilot. While they worked for their engineering and commercial qualifications respectively, they often overheard remarks like 'What do those bloody women think they are doing here?' As soon as they had their necessary licences, the girls hired a two-seater Moth and parked themselves optimistically in a field between Sevenoaks and Orpington where they hoped to attract passengers for joy-rides. They took £4 on their first day, and for Pauline's twenty-first birthday her father gave her a two-seater Spartan. The rest of the summer of 1931 was spent in a large field near Wallingford: a tumble-down farm shed sheltered the aircraft at night, while the girls lived in a single-roomed hut. Business was slow although it picked up a little when **they started accepting air taxi fares as well as giving joy-rides.**

The following spring, with a larger Spartan, a three-seater acquired during their winter hibernation and christened *Helen of Troy*, Dorothy and Pauline were invited to join an air circus called the Crimson Fleet. Their takings on the first day were £30, a considerable improvement on their previous average. By the beginning of 1933, they still had £300 to pay off on the Spartan, and toured with a bigger air circus, the British Hospitals' Air Pageant, run by a man named Barker, who had previously been company secretary to Alan Cobham. For several years, Cobham's name had been synonymous in England with aerial entertainment, through which he had, as he put it, 'taken aviation to the people'. He felt Barker's defection to be a betrayal, the more so as he regarded Barker's claim to be operating as a charity for hospitals to be deceptive. The hospitals did benefit, but so did the organiser, who, according to both Cobham and C.G. Grey, editor of the *Aeroplane*, was exploiting the gullibility of the charitably minded.

Barker was also providing entertainment and employment, although circus life was tough. Performers and mechanics slept in caravans and tents, moving to a new town every morning: before the afternoon show could start every aircraft had to be thoroughly checked. The programme began with a formation flight on which passengers were carried, so that everyone for 15 miles around the site knew that there was a circus in town. After this, joy-riding in earnest started, the discrepancy in weight between passengers often causing hilarity and sometimes even danger; at least one pilot was constantly busy taking passengers for aerobatic trips.

Balloon-bursting, paper-cutting, bottle-shooting, and height-judging, all from the air, provided circus entertainment, with a certain amount of innocent deception. For the bottle-shooting, for instance, an endless supply of empty beer bottles was lined up in front of a screen and, as the pilot flew past, he or she would throttle back and fire a pistol: the cartridges were blank, but the noise was realistic, and the bottle always shattered, as it was hit from behind the screen by a sledge-hammer. Paper-cutting involved throwing a toilet roll out of the aircraft and cutting the streamer into small pieces with the propeller and wires of the aircraft 1000 ft above the spectators. Crazy flying, in which the pilot did everything he or she should not do – apparently stalling a few feet from the ground, sweeping along sideways with one wing brushing the grass, making for some obstacle and, at the last minute, hopping over it – always proved popular; but the greatest, and final, attraction was invariably the parachutist.

As soon as the afternoon performance was over, work would start in preparation for another in the evening. At the end of the day, aircraft had to be made secure for the night, log-books filled in, the number of passengers checked, money counted, and calculations made. Then, early the next morning, the whole circus had to move on again. In one summer, Pauline and Dorothy visited 185 fields, each with its own problems and peculiarities. Autograph-hunters besieged the pilots on the ground; photographs of the aircraft in which the intrepid first-time fliers had been taken up were even more popular, with the advantage that they could be sold. There was always an element of danger, however, even on the ground: during the 1933 tour, for example, three arms were broken by propellers, one parachutist was killed when his parachute failed to open, and a twenty-two-year-old pilot died after his aircraft caught fire when he flew into a hillside in fog.

By the end of the season Pauline's Spartan had carried 6000 passengers and the bank balance was looking much healthier. The following summer, she and Dorothy decided to avoid the constant hassle of moving from site to site by basing themselves at Hunstanton, living in a gypsy caravan. Pauline did the shopping and housework while Dorothy saw to the aircraft maintenance, and by half past ten they were ready for their first passengers. As the police would not allow the noise of an aircraft over the town, they advertised their presence by flour-bombing a speedboat: the ensuing battle, entered into with the full consent of all involved, delighted even the boat passengers. Joy-riders were charged 3s 6d for 1½ minutes, rising to 5s for a circuit of the town and a short sortie over the sea, and 7s 6d and 10s for correspondingly longer flights. A return trip across the Wash between Hunstanton and Skegness cost £1. On bank holidays and in good weather during August, there were as many as 150 passengers a day, bringing in over £30.

Half-way through their first summer, a parachutist, Bill Williams, joined them, and they devised a full display of Sunday afternoon aerial entertainment, starting with crazy flying and continuing with a demonstration of wireless-controlled flight and dancing to music. Wearing earphones, Pauline would answer questions put to her by Dorothy from a loud-speaker van, answering 'Yes' by pushing the joystick backwards and forwards so that the nose of the aircraft nodded, and 'No' by tipping the wings first to one side and then to the other. When the conversation was over, Dorothy put on a gramophone record of a waltz: Pauline then performed a 'falling leaf' descent which, from the ground, looked as if it was in time to the music. The audiences never realised that the aircraft had no radio. During their second summer over 1000 people turned up every day for their air display afternoons, travelling by excursion trains from King's Lynn and Wells.

For the summer of 1936 they joined Tom Campbell Black's Air Display as chief pilot and chief engineer respectively. The season started in April with bitterly cold weather, and continued to be dogged by both bad weather and bad luck. In May Pauline was about to take off with two passengers on a demonstration air race when another pilot decided to take off in the opposite direction: the aircraft met at the centre of the field, providing unusual excitement for the crowd. The only injury was to Pauline's head, which needed several stitches and kept her grounded for nearly a month. At the end of that season, Pauline and Dorothy ended their partnership, Dorothy leaving to run an aerial garage for a year, after which she married and, during the war, worked at Farnborough. She and her husband were killed in an air crash in South America not long after the end of the war.

Pauline Gower, who had accumulated at least 2000 flying hours by the time the war started, founded and ran the women's section of the Air Transport Auxiliary (ATA). By the end of the war, the ATA had employed more than 150 British women, and nearly three times as many men, as ferry pilots, a gratifying breakthrough for women who wished to fly although they were still banned from military flying service. A few of the women who flew with the ATA had flown privately or, like Pauline, professionally; others had learnt to fly through a government-sponsored scheme shortly before the war.

From October 1938 subsidised flying was available in Britain through the Civil Air Guard (CAG): 57 out of the country's 59 flying clubs joined the scheme, which cut the rate of flying from

£2 or more to 2s 6d per hour. Within a fortnight there were 34,000 applicants, and by July 1939 10,000 prospective pilots were being trained and several thousand licences had been issued. Although women were in the minority, nearly 1000 were accepted and some 200 acquired licences. As Joan Hughes, a young instructor, put it: 'Everybody was having a go, even old ladies of seventy.' There was, however, some male opposition, openly expressed by C.G. Grey in the *Aeroplane*: 'The menace is the woman who thinks that she ought to be flying a high-speed bomber, when she really has not the intelligence to scrub the floor of a hospital.'

Twenty-year-old Joan Hughes felt that 'everything fell into one's lap' when the CAG offered subsidised flying. She had started flying when she was fifteen to keep up with her brother – her parents allowed her 40 minutes a week in the air, at £2 10s an hour – and had gained her PPL at seventeen. Since then, she had been struggling with the cost of increasing her flying time. As an instructor at Romford Flying Club she brought her hours up to 500, and had taught forty or fifty people to fly by the time the ATA was formed.

Rosemary Rees, the daughter of a multilingual civil servant in India who had retired into politics and a caricaturist from the old Catholic family of Dormer, also became a CAG instructor, with the idea of doing something useful, but she never instructed. Her wealth – although she would never admit to having been rich: eccentric perhaps, but not rich – had enabled her to own a private aircraft. Rosemary and her brother Richard were brought up in a household where the social status accorded to the local MP was taken as their due. Like many parents of the day, Rosemary's thought that education was important for a boy but not for a girl. Their son went to Eton and Cambridge, and Rosemary 'came out', but did not stay out for long. Her mother was not 'so silly as to hold a dance or treat the London season as a marriage market', but Rosemary was presented at court, curtseyed in front of King George V and Queen Mary, and did the round of London dances and dinners.

Turning her back on all the socialising, Rosemary Rees became a dancer. She never considered herself top star material, but she was small and slim, with a natural suppleness which led her into acrobatic ballet: even in her seventies she could still touch her toes. With a touring ballet company, she danced on the piers of seaside resorts and 'mucked in happily' with the tiring and far from glamorous life, although she never had to live only on her salary of £5 a week. When her colleagues discovered that her father was a baronet, they nicknamed her 'the Bloody Duchess'.

Rosemary's ballet career was interrupted when her brother's Cambridge friend Gordon Selfridge, 'son of the man who started the shop', persuaded her to share his new enthusiasm for flying. Although she could see no reason why she should want to fly, she enjoyed her first flight in 1933, and carried on to please her instructor, Captain Baker: 'I didn't really want to fly, but I didn't want to disappoint him.' All her early flying was done from Heston, the 'snob' place to fly: she was astonished when she discovered later that there were many competent pilots from other flying clubs she had never heard of, some of them even in the north of England.

Her mother bought her her first aircraft, a Miles Hawk Major which cost £800, and she later bought a second, a Miles Whitney Straight. To protect it from bird droppings, she made it a 'hangar coat'. Much of her time in the years before the war was spent touring Europe with Gordon Selfridge, in separate aircraft, landing on private airfields belonging to a wide circle of wealthy friends. At first, flying round Europe was free of red tape, but as war approached restrictions and tensions became more frequent and irksome until, in Berlin in 1938, Rosemary was advised to take her aircraft home to avoid having it taken from her.

For Veronica Volkersz, the CAG made learning to fly financially possible. Her enrolment on a flying course made newspaper headlines, 'Beauty Queen joins Civil Air Guard', although the nearest she had come to being a beauty queen was her participation in a pageant, riding side-saddle and elegantly attired in flowing blue and silver. Riding, both during her early years in India and later in England, where she regularly missed school to go hunting, and driving an Aston Martin sports car, satisfied her love of speed until she had her first flight. A racing driver with his own Gipsy Moth took her up on a sunny day in 1938 at Brooklands. The flight, complete with aerobatics, was enough to convince her that she had to learn to fly. For a total outlay of £2 10s she acquired her A licence through the CAG. With an overdraft and a bank loan she had completed half the flying hours needed for a commercial licence by the time private flying was banned at the outbreak of war.

By then, the British had drawn up pessimistic contingency plans for a reserve of civilian pilots who could be called on when enemy bombing destroyed vital services and cut communications. The Air Transport Auxiliary was given official approval on the day war broke out, but as postal, medical and emergency services continued to function it never had to fulfil its intended role. Instead, under the banker and director of British Airways Gerard d'Erlanger, commonly known as 'Pops', and Lord Beaverbrook, Minister of Aircraft Production, its role was changed to delivering RAF aircraft from factories to dispersal airfields and Air Force squadrons. The initials ATA soon gave it the nickname, sometimes derisory, sometimes affectionate, of Ancient and Tattered Airmen: young and able-bodied men were needed in the RAF.

In December 1939, in spite of initial male opposition, Pauline Gower was appointed to form a women's ferry pool at Hatfield with eight pilots, of whom Joan Hughes was the smallest and youngest and one of the most experienced was Rosemary Rees, with her 500 hours in the air. Veronica Volkersz, with far fewer flying hours, had to fill in time driving ambulances, but wrote to the ATA every month for eighteen months begging to be taken on: her impatience was rewarded in March 1941. Six months later, she met Gerry Volkersz, a tall, good-looking lieutenant in the Royal Netherlands Naval Air Service, who had escaped from Holland and was attached to the RAF as a Spitfire fighter pilot: they were married in November 1942.

Veronica's was by no means the only ATA romance. In 1944, Diana Barnato Walker, one of the most glamorous of the women pilots, married a young RAF pilot, Derek Walker, who would, she was convinced, have 'got to the top' if he had not been killed soon after the war in a crash at Northolt. Like Rosemary Rees, Diana had been presented at court, in 1936, wearing a Norman Hartnell dress, and was much photographed at balls, races and ski resorts. Her beauty, remarkable even in her sixties, was carefully preserved at great expense after a teenage riding accident in which her face was badly injured: far from deterring her from riding, she was later Master of the local hunt for thirteen years. She started flying to escape her governess and the restrictions of her wealthy socialite upbringing as the daughter of the millionaire racing driver Babe Barnato. Her grandfather, a penniless East End Jewish juggler who became one of the richest and most influential men in the South African diamond boom, was, according to Diana, murdered by his nephew and business associate: officially, he jumped overboard near Madagascar on his way from South Africa to England in 1897.

When she joined the ATA in December 1941, Diana had been driving ambulances for the Red Cross and had not flown since the start of the war. She had done only ten hours' solo and, before the admission test, practised with a boyfriend in the RAF – on a sofa which they 'flew' as if it was a Tiger Moth. In the ATA, she showed an occasional reckless disregard for such rules as the ban on aerobatics. When she succumbed to the temptation in a Spitfire, her face powder fell out of her pocket while she was upside down, and scattered over the cockpit. After she landed, with the evidence still smearing the Spitfire's perspex canopy, the reaction of the ground crew was to say, 'Well, there you are, what can you expect when you get women flying aeroplanes?' She was caught out too on another occasion when she flew unofficially between delivery trips to have lunch with a friend in Cornwall, and was demoted by being sent from White Waltham, the ATA headquarters, to the women's pool at Hamble: an ineffective punishment, as it turned out, for she preferred Hamble, where she shared a cottage with another woman pilot.

The women pilots, who were initially paid £80 a year less than men doing the same job, inevitably attracted considerable publicity. They were at first accused of 'encroachment on men's jobs', of 'not seeking this job for the sake of doing something for their country but for the sake of publicity', 'doing it more or less as a hobby', and 'getting paid for drinking tea'. They were also praised. When women had been ferrying aircraft for nearly two years, a publicity film was made about a day in the life of a woman ferry pilot. Veronica Volkersz was chosen as the pilot, and shown making an urgent delivery of a Hurricane to a fighter squadron: pressing on madly through storm and tempest, on arrival she found she could not get the undercarriage down, and eventually, game to the last, made a brave belly landing, stepping nonchalantly out of the wreckage. To the general public, it may have been an inspiration to see what Britain's brave women pilots were enduring for the sake of their country. Veronica herself found it corny, embarrassing and inaccurate: the studio mock-up of a Hurricane had a Spitfire canopy, the smoke machine which was to create the effect of the bad weather in which she was to register anxiety, merely made her cough, and the dialogue bore little relation to reality.

In the first winter of the war, the women ferried bright yellow open single-engined Tiger Moths to the north of England and Scotland, making their own way back, often overnight in crowded trains: some resorted to sleeping in the luggage racks. The Tigers were cold and draughty, and it was a bitter winter. Rosemary Rees, being thin, suffered particularly from the cold. Even in less draughty aircraft, she wrapped herself up in as many clothes as possible and always took a fur cushion to sit on.

Women eventually flew every possible aircraft except sea-planes, from which they were banned because of the embarrass-ment which it was thought would be caused if they were stranded afloat overnight with male crews. Aircraft were divided – according to size, power and difficulty – into five groups, and pilots were checked out on one aircraft in each category. Bob Morgan, chief technical officer for the ATA, was given the task of preparing concise information for each aircraft, and with the help of these invaluable notes pilots were able to handle all other aircraft in the group. It was not unusual for a pilot to read the Ferry Pilots' Notes about an unfamiliar aircraft while taxiing before taking off or even in the air, a habit which caused con-siderable alarm to male ATA and RAF passengers, who were at first often appalled at the idea of being piloted by women.

The first woman to fly the fifth group, four-engined bombers, was Lettice Curtis, who had joined the ATA reluctantly. She began flying after taking a mathematics degree at Oxford, and although her friends had warned her, 'You'll never get a job', was taken on as a junior pilot for a firm carrying out Ordnance Survey work. When she was grounded at the outbreak of war, she became an Ordnance Survey researcher, but kept getting letters from Pauline Gower: in the summer of 1940 she 'fell'. In spite of her misgivings, Lettice, whose wartime passengers included a young flight engineer called Freddie Laker, found the ATA 'a marvellous experience'.

Only ten other women, including Joan Hughes and Rosemary

Rees, qualified to fly the four-engined aircraft. Rosemary was at first amazed at her ability to fly so many different aircraft, but then decided it was a question of state of mind, 'like doing cartwheels'.

Even the more advanced aircraft, with the comfort of closed cockpits, had no radio, and the pilots had to cope unaided with whatever hazards they might encounter. Barrage balloons were almost as much of a problem to ferry pilots as they were to the Germans, for whom they had of course also to be constantly on the lookout. The ATA women might be called on to ferry several different aircraft in one day, or take several days over one long-distance delivery. Weather was the chief enemy, with the frustration of waiting and the determination to avoid delays often making it difficult to accept being grounded.

Fifteen women were killed flying for the ATA, among them Amy Johnson, and many others had lucky escapes. When First Officer Diana Ramsay crash-landed at White Waltham in a Tempest, the ambulance and crash crew were surprised to find her sitting on the wreckage, terrified of the cows in the field. Another woman pilot came down in the sea off the east coast of Scotland in a Barracuda, which submerged to the seabed: she somehow managed to wriggle free and float to the surface, and was rescued by the crew of a fishing smack who, she claimed, nearly threw her in again when they discovered her sex. Diana Barnato Walker was flying back from Brussels in formation with her husband – she had been given special permission for the foreign trip, a belated honeymoon – when she lost her way in thick fog, and was lucky to land safely at Tangmere, the only airfield open in the fog-bound south of England: she made a perfect landing, commenting that 'You always do when you're scared'. Rosemary Rees had a close shave with Blackpool Tower when she was flying under low cloud to Scotland: 'Suddenly something flashed by me. "Pfft!" I thought, "Good God, Blackpool Tower". I had quite forgotten it. It would have been a ridiculous (and to other people rather amusing) way to crash oneself.'

The ATA eventually expanded until it had twenty-two bases, including two women's pools. A Leicestershire MP, Sir Lindsay Everard, the chairman of the Royal Aero Club and an Honorary Air Commodore, turned over his private airstrip to No 6 Ferry Pool at Ratcliffe. Several pilots, both male and female, lived in his manorial home, including some of the twenty-six American women who joined the ATA during the last two years of the war: one of them, Mary Ford, was even married from Ratcliffe Hall. The pilots dined at a long table with Sir Lindsay and Lady Everard, served by a butler known only as Smart who once introduced Air Vice Marshal Popham as 'Air Popham'. The private swimming pool was used for ditching practice.

By the middle of 1945, the ATA was gradually being run down. It had grown from its hastily improvised beginning into a highly complex organisation, with a ground staff of more than 2500 men and women. Its pilots – more than 700 altogether of whom 54 were killed – had flown 750,000 hours with 147 different types of aircraft. The organisation was officially closed in September 1945, with an Air Pageant opened by Lord Beaverbrook, but it lingered on at White Waltham until November. Pauline Gower, who died giving birth to twins not long after the war, and Margot Gore, Commanding Officer of the women's ferry pool at Hamble, were awarded the MBE, as were Joan Hughes and Rosemary Rees.

For the women ferry pilots, victory meant the end of an era which had, for all the misery inflicted elsewhere by war, enabled them for a few years to fly professionally. They had enjoyed the spirit of camaraderie, the sense of professional pride, and the sheer pleasure of flying aircraft which they would otherwise never have been able to fly.

More women were employed as pilots in Britain during the war than at any time before or since, although even then they were excluded from flying in combat. The only woman to have flown on active combat before 1939 was Sabiha Gokçen, a Turkish orphan adopted by Kemal Ataturk: she piloted one of nine aircraft which, in 1937, bombed the Kurds for a month.

Some of Germany's most successful wartime test pilots were women, although they kept their civilian status. It was only in Russia that women were allowed to fly on combat missions, making night attacks over the German lines in slow old PO2 biplanes with open cockpits in the bitterly cold Russian winter: each aircraft carried only four bombs and had to contend with anti-aircraft fire, but between them the women apparently dropped 23,000 tons of bombs. In France, women were allowed to act as ferry pilots for a few weeks, but were then told instead to serve by knitting for their country. In spite of their experience and proven ability, few European women managed to find post-war employment in aviation.

Flying had become such a habit for Rosemary Rees that, after the war, she set up her own small flying business, Sky Taxi, with an old Proctor and one taxi-pilot: herself. To do this, she had to get her commercial licence: no problem from the flying point of view, but she had always had to 'cheat' to pass medically because of her short sight. She had learnt the eye test card by heart before taking her PPL, and at every subsequent medical examination had recited it with suitably convincing pauses. The chart used was, to her dismay, changed before the medical for her commercial licence. Dressed up as an old lady, she made her way into a doctor's room in Wigmore Street while its occupant was at lunch, and copied the new test. She was still only granted her commercial licence provided 'correction to defective vision' was worn, and was given glasses: she never used them for flying, as by then she was so used to the handicap of her short-sighted astigmatism that she did not want to adjust to perfect vision.

She ran Sky Taxi for four years until soon after her marriage to Philip du Cros, whose family had known hers since the First World War and who was, like her father, a Conservative MP, in Devon. There were, however, not enough customers in Bideford and Barnstaple to make a flying taxi service viable, and it was with some relief that Rosemary gave up her Sky Taxi to immerse herself in local politics. She never flew again, nor wanted to: she hated the idea of radio and controllers organising her in the air. Nor did she wish to have a family: 'I'm not a sexy breeding type, I'd have hated everything about it – I'd have hated the pain of the birth, and I dislike children.'

Although Barnstaple proved the end of Rosemary Rees' flying career, for Ann Welch its small grass field had, five years before the start of the war, been a starting point. Ever since her first flight with Alan Cobham's Air Circus in 1930 when she was thirteen, Ann had longed to fly. She had kept a record in an Aeronautical Society Diary for 1933 of every aircraft she saw that year, complete with the weather conditions in which she had observed it, and read its eighty-two pages of aviation information avidly. Croydon and Biggin Hill airfields were within cycling distance of her home, and she cycled first to Biggin Hill, where the RAF commanding officer allowed her to sit in the cockpits of

Bristol Bulldogs and Hawker Demons, and then, whenever she had 5s – the cost of a flight – to Croydon. When she was sixteen, she bought a motorcycle for £5 to make her visits to Biggin Hill and Croydon easier – and rode it, unlicensed and uninsured, to school, which as a result she soon left with the terse report from her headmistress that 'Ann has been with us ten years'.

Her parents, alarmed at her obsession with flying, tried diversionary tactics: she was sent to stay with the artist Charles Tunnicliffe and his wife to learn how to paint, but was delighted that his love of birds was in a way a love of flying, and that the Lancashire Aero Club at Woodford was within cycling distance. For her seventeenth birthday, her father gave her £30 towards her A licence, which she spent during a family camping holiday in north Devon above what is now Chivenor airfield but was then Barnstaple aerodrome. By the end of the holiday, she had, in September 1934, flown solo, but because of bad weather had not had time to gain her licence, which she was awarded a month later at Brooklands.

During the next two years, Ann flew whenever she could afford to, and took part in what were known as 'breakfast patrols': two would share the cost of a Brooklands' club aircraft, one being pilot on the way to another airfield under friendly attack and the other flying home. In the middle of 1937 when she had logged 50 hours' flying, she joined a gliding course at Dunstable run by an Anglo-German Fellowship. There were sixteen young Germans and nine British on the course: Ann was the only girl. Gliding in Britain was, as she put it, regarded by most pilots as 'aerial tobogganing and not real flying at all', but any chance to be in the air was worth taking, and she relished the 'unexpected, the new, the beginning' of the small and closely knit gliding fraternity. Gliding had only recently been introduced to Britain from Germany, where un-powered flight had been developed during the post-war years when Germany was not allowed to have an air force.

By the time the Second World War began Ann was a glider instructor and had started the Surrey Gliding Club. When the ATA was formed, she applied, and in 1940 was accepted, although she could offer only 100 hours' flying experience in powered aircraft and gliders combined. 'The war', she said, 'was just beginning to get serious and I had to be involved; and it had to be in flying. Nothing else could even be contemplated.' In her two years with the ATA, Ann, like the other women ferry pilots, enjoyed the varied flying and the company, and survived a number of 'close calls'. She left the ATA in 1942 to start a family.

After the war, Ann Welch returned to gliding, and was Manager of the British Team at World Gliding Championships for two decades from 1948, being awarded first the MBE and then the OBE for services to gliding; and later the Lilienthal Medal in 1974 and the FAI Gold Air Medal in 1981. She was also prominently involved in the development of hang gliding and microlights, and wrote a number of books on gliding and hang gliding as well as her autobiography, *Happy to Fly*. With her second husband she toured to World Gliding Championships, at the same time bringing up three children; and as a grandmother she developed her interest in sailing both on a family yacht and as a foredeck hand on the three-masted schooner *Sir Winston Churchill* – the first grandmother in the crew.

For Joan Hughes, the ATA and flying had been an alternative to marriage. 'You can't join the ATA,' her fiancé had told her, 'you're going to marry me.' That she remained single seemed to her in retrospect a reprieve from the horror of 'being trapped by domesticity'. Flying was the most important thing in her life – she would, she claimed, have gone without food rather than without flying – and after the war once again 'it all fell into one's lap'. From its start in 1947 until 1961, Joan was an instructor at the West London Aero Club at White Waltham, and was then employed by British Airways Flying Club, still at White Waltham, moving with it in 1966 to Booker, later known as Wycombe Air Park, where she continued to instruct instructors on part-time courses after her retirement in 1982.

As an enjoyable change from instructing, Joan participated as a pilot in two films, *Those Magnificent Men in their Flying Machines* and *The Blue Max*. In the former she flew a replica of the dainty little Demoiselle designed and flown in Paris in 1910 by the diminutive Brazilian, Parisian Santos-Dumont, whose weight was the same as Joan's. With her eight stone on board, it would fly; with another stone or two – in other words, with an average man – it would not. The pilot's position, sitting on the undercarriage, was both unusual and vulnerable, but Joan found it a delight to fly in calm conditions, although alarming in wind and very sensitive to her position.

Diana Barnato Walker continued to fly, giving the members of the Girls' Venture Corps flying experience until she needed glasses in the late 1970s. On a course at Middleton St George, a casual remark by a Training Squadron pilot gave her the idea of flying a Lightning jet fighter. As she was socially acquainted with the Air Minister, Sir Hugh Fraser, she made an official appointment to see him, and gave him 'twelve phoney reasons' why she should be allowed to do so. Sir Hugh, not averse to the publicity which the venture would attract, gave his official permission. After 30 hours of training on a simulator, Diana flew 10 miles high at 1250 mph on a sunny July day in 1963, joining the high-speed élite.

Friendships made in the ATA were maintained after the war through an ATA Association Newsletter and regular reunions. In 1983 a party of British members travelled to America for the dedication of an ATA Museum: there is no British equivalent to preserve the memory of the indigenous ATA pilots, male or female. Ann Wood was hostess at a reunion near Boston with many of her British wartime colleagues. She had started flying in the summer of 1940 on a free civilian pilot training scheme initiated when America became aware of the 'oncoming war'. Fifteen boys were needed for a flying course on which her brother had enrolled, and in the absence of a fifteenth male applicant Ann was accepted. When the first contingent of American women pilots joined Britain's ATA, she was 50 hours short of the minimum requirement of 350 hours, but was accepted later. After the war, she held a number of ground-based public relations positions both with Boston-based aviation companies and with Pan-American Airways, for whom she eventually became Vice-President of International Airport Charges in Washington and New York.

Although women had proved that they could fly as competently as men, after the war they still met active discrimination when they applied for flying jobs. Lettice Curtis wanted to be a test pilot, signing her application as E.L. Curtis and giving no clue about her sex other than Benenden School and St Hilda's College, Oxford, on her curriculum vitae. She was called for an interview at Farnborough, and was sitting in the waiting room when a messenger was sent in to tell her: 'This is the board for test pilots.' 'I know,' she replied, and a few minutes later heard gales of laughter from the boardroom, where the idea of a woman applying to be a test pilot was clearly a source of amusement. There were several hundred ex-RAF applicants, but she was nevertheless put at the top of a reserve list, and later in the year was invited for a flight test.

She was offered a job, but then 'all hell broke loose'. The Chief Superintendent and the Senior Controller of Aircraft were both adamant: no women. Lettice had to admit defeat, and worked for five years on the technical staff at Boscombe Down, where all civil aircraft were sent for their Certificate of Airworthiness tests, then as a flight development engineer on Fairey's Gannet at White Waltham, with an experimental air traffic control unit at Heathrow, and as a contract engineer for Sperry's at Bracknell, where she was involved with the early development of Polaris. As the author of a meticulously detailed book *The Forgotten Pilots*, Lettice Curtis became the accepted expert on the women's contingent of the ATA, of which she had initially been such an unwilling member.

When the Royal Air Force Volunteer Reserve opened its ranks to women after the war, in 1948, Veronica Volkersz, like several other ex-ATA women, became a reserve pilot. Her marriage had failed, and she was faced with the problem of earning her living: flying had to be the answer, and she started to work again for her commercial licence. Although many of the women volunteers had considerably more flying experience than the male reservists, the RAF treated them as beginners. Veronica found herself back to flying Tiger Moths, 'better than nothing', but gained experience in aerobatic, instrument, night and formation flying, all new to her.

Even with her combined ATA and RAFVR experience, as well as a commercial licence, Veronica's efforts to find employment resulted in a large file of rejections on the grounds that 'It is not the policy of the company to employ women pilots', and she too was often able to obtain interviews only by signing her initials rather than her full name. Her first flying job was as a ferry pilot delivering Tempests to Pakistan for a firm called Mayfair Air Services. Although she had flown Tempests in the war, she had not done so since, and had to be checked out at the Hawker factory at Langley: the men gathered to watch as she took off, but she was refused permission to join the test pilots in the mess because 'the chaps like to swear a bit with their lunch, and if there's a woman around, it rather puts them off'.

She flew five delivery trips and added 90 hours to her flying experience on Tempests. When the contract ended and another, for 50 Furies, went to a rival firm, all the male pilots who had flown with her were taken on, but the company refused to employ women, although some of the 'ex-RAF types' engaged to fly the Furies had neither flown a fighter before, nor knew the route.

Her next employer was a woman. Monique Agazarian, another ex-ATA pilot, was managing director of Island Air Services, and needed a pilot for joy-riding and charter work when her husband left the firm to join BEA. It was Festival of Britain year, 1951, and for £1 10s a trip Veronica flew joy-riders over the Festival site in a Rapide, inflicting a surcharge of 2s 6d on any passenger who was airsick. After a summer of joy-riding and charter work, she applied to the Civilian Anti-Aircraft Cooperation Units (CAACU) in Norfolk for a job as a civilian pilot for an Air Ministry contract with RAF aircraft. She could fulfil all the requirements: membership of the Reserve or Auxiliary Forces, previous experience on Spitfires, Beaufighters and Oxfords, and an instrument rating. The authorities agreed reluctantly that she could be taken on: they had given the matter 'a lot of thought', had done their best 'to find a reason to withhold approval', but had to agree that 'the lady fits all the qualifications'. The flying involved Spitfire low-level beat-ups and dive-bombing attacks on gun positions, target towing with Beaufighters and night-time searchlight practice which, because of the blinding effect of the beams, meant long stints of instrument flying. Once a week, there was a 'maximum effort': starting at 8.30 a.m., there would be four or five hours' flying by day, followed by two hours at night and another hour at dawn. Veronica finished by buzzing the guest house where she was living to give warning that she would soon be ready for breakfast.

In 1955, Veronica, who was then flying Oxfords with a Fighter Control Unit at Rochester, was invited to take part in a BBC television *Find the Link* panel game as one of two women who had flown jets: the other, Mary Wilkins, was also an ex-ATA pilot. Veronica's name had been suggested by the British Women Pilots' Association which had just been formed 'to promote the training and employment of women in aviation'. The proven ability of women as wartime ferry pilots had been conveniently forgotten by the majority of the men who, ten years later, still dominated British aviation and felt that women had no professional place in the air. Only seven of the ATA women pilots were still flying commercially in 1955, and only two others had acquired commercial licences.

Equality in the air was still a long way off.

❮❮ 13 ❯❯

Hanna Reitsch:
Patriot

Hanna Reitsch was a well-known glider pilot before the Second World War, and felt that there was no better instrument for peace than her 'beloved gliding'. She nevertheless spent the war years putting her expertise at the disposal of the Luftwaffe, as one of Hitler's most famous test pilots. As a prisoner after the war, her offence was, she said, that she was 'a German, well known as an airwoman and as one who cherished an ardent love of her country and had done her duty to the last'.

Born in 1912, she was brought up to be a good German wife and mother, but from the age of thirteen was determined to become a flying missionary doctor: her father, an eye specialist, approved of her interest in medicine, but failed to prevent her learning to fly. On a gliding course at Grünau, she was the only girl: just over 5ft and weighing only about 90lb, she found it initially difficult to be taken seriously. Soon, however, she had broken endurance and altitude records, both unintentionally, the first by staying up for five hours, the second by allowing herself to be drawn up through a storm cloud. Wearing a summer dress and sandals, the initial confidence of inexperience was soon 'dissolved and submerged in fear' as the glider was buffeted violently by wind, rain and hail: she briefly lost control when ice froze the instruments.

Hanna's initiation into competition gliding, at the 1933 Rhön Soaring Contests, was however a dismal failure: time after time her Grünau Baby flopped miserably down to earth as faster and more streamlined competition gliders soared effortlessly. The Wasserkuppe beside the Rhön, where the annual Soaring Contests were held, was the birthplace of gliding: from 1920 onwards, German enthusiasts, frustrated by the Versailles Treaty ban on powered flight, gathered there to experiment with engineless flight, at first lifting their rudimentary gliders – which were sometimes repaired wartime aircraft with the engines removed, sometimes improvisations – for only a few seconds.

Through her determination Hanna won a place on an expedition to study thermal conditions in South America, which she financed by standing in for the star of a film in gliding sequences. During the three-month tour the glider pilots often followed South American black vultures to find thermal air currents, and took four of the birds home with them: they refused to fly, although one eventually left its open cage, on foot, and was later seen walking in the streets of Heidelberg; the other three were given to Frankfurt Zoo.

Although Hanna had started a university course, when she was invited to join the Glider Research Institute at Darmstadt she dropped her medical studies, and within a few weeks set a women's world record in long-distance soaring with a flight of over 100 miles. During the next few years she gained an international reputation as a glider pilot, representing Germany in Finland, Portugal, Hungary, America, Libya and Yugoslavia. In 1937, she was one of five German pilots who, for the first time, crossed the Alps by glider in a Sperber Junior specially adapted to fit her: once in the pilot's seat she could hardly move and felt as if the wings grew out of her shoulders.

As a test pilot for the institute, she tried innovatory gliders such as Hans Jacob's Sea Eagle, the first glider-seaplane; but when she attempted to launch it behind a motor boat she finished under water. Dive brakes were a new invention, and after Hanna had demonstrated on gliders that they successfully limited speed and increased stability even in a steep dive, they were fitted to military aircraft: Hitler himself made her an honorary *Flugkapitän*, the first time such an honour had been bestowed either on a woman or on a research pilot and a title which she used proudly.

By this time Germany was busily rebuilding its air force. Gliding had played an important part in the training at youth camps of potential air force pilots, who were recruited for the re-formed Luftwaffe. Hanna, who had acquired licences and experience on powered aircraft, was summoned in September 1937 by General Udet to the Luftwaffe Testing Station at Rechlin, where she was to test military aircraft, although technically she remained a civilian. Her interpretation of Germany's rearmament was rather different from that of the rest of the world: 'We young men and women wanted peace – but a just peace, which allowed people to live. And the German people wanted it . . . ' She called Germany's fighter and bomber aircraft 'Guardians at the portals of Peace', and was proud that, through the caution and thoroughness of her flying tests, she was contributing to the protection of the country in which she had such a strong faith.

Hanna's self-disciplined approach, fanatically patriotic sense of duty, and strong religious convictions had been instilled by her mother, who wrote almost every day warning against false pride and entrusting her to God. In the air Hanna combined a positive and almost mystical pleasure in flying with a need for absolute order and correctness. When, for example, she was invited to fly a helicopter for the first time, she prepared herself by learning in detail the theory of a completely new concept of flying, and within minutes was able to control the revolutionary vertical take-off, hover at will, manoeuvre sideways, backwards, forwards, and drop almost vertically to land on precisely the spot from which she had taken off.

At the beginning of 1938, Hanna and the helicopter featured among the attractions of a variety programme, based on the theme of Germany's lost colonies, in the covered stadium of the Berlin Deutschlandhalle. General Udet, disappointed by the lack of international excitement over the German invention – few people elsewhere were in a mood to share his pride in Germany's technological achievements – was certain that the annual International Motor Show in Berlin would ensure the presence of the world's press. Every evening for three weeks, Hanna followed a string of dancing girls, fakirs, clowns and blackamoors with a searchlit display of the new flying wonder: *Deutschland* was clearly painted on the helicopter's silver fuselage, and she finished her act with a Nazi salute.

Among the peace-time projects on which Hanna worked was a glider capable of carrying freight, designed to be towed by a mail plane and land at places where there was no regular airborne postal service. When her painstaking tests had proved that such a glider was feasible, and moreover capable of carrying passengers, the High Command of the army asked her to test a glider which could transport ten men and an officer. In spite of the success of the operation, there were dissenters who claimed that parachuting was a better way of landing troops quietly and efficiently. Even so, in a contest between ten gliders carrying infantrymen and ten military aircraft with paratroopers, the troop-carrying glider proved that it could be a valuable weapon.

As soon as war broke out, a unit of glider pilots was formed for the invasion of France. When this was postponed until February 1940, it was Hanna who tested the brakes which had been hastily designed to bring the gliders to a halt if they landed on ice: at first, she was winded by the violence with which they stopped the glider, and they had to be modified and thoroughly tested again before she considered them satisfactory.

Glider assault troops, with ten-man gliders, were successfully used in 1940 to capture a Belgian fort, and a larger version, capable of carrying twenty-one men, was satisfactorily tested. This was dwarfed by the Gigant, a glider nearly the size of a modern jumbo jet, designed by Messerschmitt to carry a tank or 200 armed men: it was intended to use 200 of the monster gliders in Operation SEA LION, the invasion of Britain planned for 1940. Hanna made a test flight on the prototype, towed by a four-engined Junkers. Take-off was assisted by hydrogen peroxide rockets which were ignited half-way down the mile-long runway, and the two-ton undercarriage was jettisoned as soon as the Gigant was airborne. She was not impressed: the towing aircraft was underpowered and the Gigant itself primitively built and difficult to fly.

When she failed to convince Udet and Messerschmitt to drop the idea, three twin-engined Messerschmitt 110s, with over 8000 hp, were used to tow the monster into the air. The take-off demanded skilled co-ordination between the three aircraft, which even with the boost of the glider's rockets were constantly on the verge of stalling. After a narrow escape when one of the aircraft failed to take off, Hanna refused to continue the tests. She was proved right when the rockets on one wing of the Gigant failed to ignite: the three towing aircraft were dragged together and 129 glider troops were killed.

A flying petrol tanker was another unworkable project tested by Hanna, the only pilot small and light enough for the glider which was eventually to be towed pilotless as an aerial refuelling tank: she struggled with nausea and fear as it twisted and turned behind the towing aircraft.

The idea of using a series of ropes to restrain an aircraft landing on a ship's deck met with equally little success and was abandoned after three trials, during one of which Hanna was almost decapitated by a rope. During more successful experiments with a device to cut the cables of the barrage balloons which were frustrating German aircraft over London, Udet was a witness to another of Hanna's lucky escapes when a balloon cable parted and shaved off the lower edges of two propeller blades.

Hanna Reitsch met her Führer again in March 1941 when, after the cable-cutting accident, he conferred on her the Iron Cross – Second Class. The previous day, she had for the first time been introduced to Goering, who commented on her diminutive size and awarded her a Gold Medal for flying '*mit Brillanten*'. Not to be outdone, her home town of Hirschberg summoned her to receive the Freedom of the City and a Grünau glider.

Her most dangerous project was the Messerschmitt 163 experimental rocket plane. To fly this was 'to live through a fantasy of Münchhausen': 'One took off with a roar and a sheet of flame, then shot steeply upwards to find oneself the next moment in the heart of the empyrean.' It reached a speed of over 500 mph in a few seconds and a height of 30,000 ft in one-and-a-half minutes, climbing at an angle of 60°–70°. Even on the ground, it was all Hanna could do to hold onto it as it was rocked by a series of explosions, but before it could be used to intercept, split up and attack enemy bombers, it had to be fully tested. The first four tests were successful. On the fifth the undercarriage jammed: it was supposed to be jettisoned soon after leaving the ground, at a speed of 250 mph. As it was unthinkable to waste such a valuable machine and its instruments by baling out, Hanna attempted to land, lost control during the high-speed approach, and crashed.

Her initial reaction was one of relief that she had landed the right way up. Then, noticing blood streaming from her head, although she felt no pain, she put up her hand. There was nothing but an open cleft where her nose should have been, and every time she breathed bubbles of air and blood formed along the edge of the cleft. Before passing out, she took a pencil and pad from her pocket and sketched what had happened before the crash. She was so seriously injured that even after the attention of an excellent brain surgeon she was not expected to live, let alone fly again. For five months she hovered in hospital between life and death.

When she was released from hospital with a reconstructed nose, she was suffering from constant headaches and giddiness, but travelled, alone, to the isolated summer home of some friends, where she organised her own convalescence. After roof and tree-climbing exercises, mountain walking, and concentrated retraining of her mind for co-ordinated thought, she started flying again in secret, and after a few weeks declared herself fit to resume normal duties.

Soon after her recovery, Hanna visited Goering, and was horrified to discover that he believed that the Messerschmitt 163 was already in mass production. 'It became,' she later wrote, 'only too clear that Goering did not wish his comforting illusions to be disturbed.' In spite of official propaganda, according to which victory was still just round the corner, Hanna knew that the situation was bleak and that German airpower was danger-ously diminished. Colonel-General Ritter von Greim, for whom she had a great admiration – they may have been lovers, as was commonly assumed – was attempting to supply air support for his sector on the hard-pressed Eastern Front, with an inadequate supply of aircraft. To sustain the morale of his men, he invited Hanna to visit them on the front line. With typical thoroughness she set about dispelling her fear by learning to distinguish between the sound of German and enemy shells, but when the men asked her about the progress of the war she 'tried hard not to raise false hopes'.

Hanna claimed resolutely that gas-chambers were no more than an invention of enemy propaganda, and confronted Himmler, whom she had previously feared and loathed as 'an adversary of Christianity', but who had redeemed himself by sending her small gifts and get-well notes while she was in hospital. Although he neither denied nor confirmed the rumours, Himmler gave her the impression that he shared her indignation at the slur on Germany's honour.

It was in an attempt to bring the war to a swift and successful end that Hanna Reitsch became involved with the development of what was to be Germany's ultimate weapon: the suicide rocket, a development of the 'buzz bomb', the V1 pilotless rocket. From the middle of 1943 it was becoming increasingly obvious that, as the death toll mounted, the country's resources were being depleted, and that a miracle was needed to bring a decisive change in Germany's favour. The suicide rocket was to provide the miracle. Hanna Reitsch became one of a small group of people dedicated to working on the aircraft in which volunteers would sacrifice their lives to save their country. In the winter of 1943–4, a conference attended by experts on explosives, torpedoes, navigation, radio, maritime engineering, and aircraft design, and by representatives of fighter and bomber squadrons, agreed that the idea was sound. There was no shortage of volunteers prepared to be sacrificed for the good of the cause. It remained only to convince those in high command.

The opportunity to put the proposal directly to the Führer arose when, in February 1944, he summoned Hanna to receive the Iron Cross First Class for her work, and her accident, with the experimental rocket. Hitler refused to admit that the war had reached a stage when a suicide mission was either necessary or acceptable: it would be an admission of failure, and he would not admit that defeat was possible. For someone who had given so much time, skill and energy, and nearly her life, to her country and its leader, 'the appalling implications' of her realisation that Hitler 'was living in some remote and nebulous world of his own', and was apparently under a misconception about the danger into which he had led his country, was a shock. Nevertheless, Hanna seized on the Führer's statement that he alone would decide when, and if, to use the last resort of the suicide mission. So that the suicide attacks could then be started immediately, she obtained permission to develop the idea, promising not to bother Hitler further during the development stage.

Volunteers were required to sign a declaration: 'I hereby voluntarily apply to be enrolled in the suicide group as a pilot of a human glider-bomb. I fully understand that employment in this capacity will entail my own death.' Tests were initially carried out not on the V1, but on a single-seater glider-bomb, the Messerschmitt 328, which would be launched in the air from a powered bomber, and in April 1944 an aircraft factory in Thuringia was given a massive order. The factory was then destroyed in an air raid. In the meantime the V1 project had been abandoned, but it was resurrected when Otto Skorzeny, a pilot whose chief claim to fame was his aerial rescue of Mussolini from a mountain hotel-prison, took over. In less than one week the pilotless V1 was converted in a secret underground workshop to carry a pilot. A dual control training model was created, so that only four or five specially trained pilots would ever have to cope with landing. The final version was the suicide bomber itself, which, once launched, could not be landed.

Hanna's offer to test the prototype was refused, because of her civilian status, until two military pilots had crashed. Of the six male pilots who flew in the tests, two were killed and four seriously injured. By the time the project was ready to be put into operation it was too late: the Allied invasion of France had begun, and the war was all but over. That the mission had not succeeded Hanna Reitsch attributed to 'the total failure on the part of higher authority to appreciate that the suicide group was no stunt, but a collection of brave, clear-headed and intelligent Germans who seriously believed, after careful thought and calculation, that by sacrificing their own lives they might save many times that number of their fellow countrymen and ensure some kind of future for their children'. Himmler had suggested that the suicide pilots should be neurotics, criminals, and the incurably diseased, who could thereby redeem their honour, and Goebbels had extolled the virtues of self-sacrificing heroism, treating the project as a propaganda exercise.

At the beginning of 1945, when Hanna was flying dispatches between Berlin and Breslau, she paid what she realised would be her last visit to her home town of Hirschberg in Silesia. Her parents, her sister, and her sister's three children, had already fled from the Russians to Salzburg. On 25 April 1945 she was given her last wartime mission: to fly Colonel-General von Greim into Berlin, which was already surrounded. Russian troops were even positioned in the city. The helicopter she was to use was destroyed in an air attack, and Hanna flew as far as Gatow, the only airfield still open near Berlin, squeezed into the rear section of the fuselage of a Focke-Wulf 190. It was flown by the Luftwaffe pilot who, a few days earlier, had flown Albert Speer into and out of the besieged city.

At Gatow, she took the rear seat which had been added to the single-seater Focke-Wulf, behind von Greim. As he had experience of flying under fire, he took the controls as they flew into the heart of Berlin, between enemy fighters and over Russian tanks and soldiers, through 'the very fires of hell' and a barrage of shots from both above and below. When von Greim was hit in the foot, Hanna leant over his semi-conscious body to take over the flying. She managed to land near the Brandenburg Gate, although the aircraft was hit more than once. The imposing buildings which had once lined the famous Berlin streets through which she and the injured von Greim were driven to the Chancellery had been reduced to charred rubble under a pall of smoke.

Hitler had summoned von Greim to promote him to Field-Marshal and Commander-in-Chief of Germany's virtually non-existent air force: Goering had been arrested as a traitor and stripped of all his offices. Von Greim's loyalty to his leader and 'conception of honour' were, according to the adoring Hanna, 'entirely immutable and selfless', and he and Hanna uncom-plainingly joined the ill-assorted company in the Bunker where Eva Braun, Hitler and the Goebbels family were living under-ground. As the tension mounted and artillery fire above the Bunker became almost continuous, Goebbels' six children were pacified by the explanation that 'Uncle Führer' was busy con-quering his enemies and that the end was near.

The end was indeed near. Even Hitler realised that there was no hope of a change in Germany's fortunes: he summoned Hanna and gave her two phials of poison, one for herself and one for von Greim, so that they would have 'the freedom of choice', and told her that, unless Berlin was relieved, he and Eva Braun would take their own lives. Hanna and von Greim left the Bunker on 28 April, and during the return flight to Rechlin somehow managed to avoid detection. From the air Berlin, a 'sea of flame', looked as Hanna had imagined the end of the world: ' . . . below we knew there were millions of men, women, children, old people, wounded, and we were flying out of this Hell, powerless to help the others!'

From Rechlin, Hanna flew her hero via Plön and Dobbin to Lübeck, where on 30 April they heard of Hitler's death. Not only had Hitler died: he and Eva Braun had married, then carried out their suicide pact, and Frau Goebbels had killed all her children before she and her husband took their own lives. At the same time Hanna learnt that all her family except her brother Kurt was also dead. It was rumoured that all refugees would be sent back to their place of origin, which for the Reitsch family would have meant living in Russian-occupied territory. Dr Reitsch considered this a fate worse than death, and took upon himself 'the heaviest responsibility of all' with a mass family suicide.

Hanna herself entered a suicide pact with von Greim: she would use her phial of poison eight days after his death, to prevent enemy propaganda from dishonouring him by linking their names. Although von Greim used his share of the poison, before the eight days were up she had decided that she must break her word to 'one of the greatest and noblest officers of the German Wehrmacht' for the greater cause of Germany, honour and truth.

Like many other leading Germans with scientific or military expertise, Hanna Reitsch was approached by the Americans: with her wide experience of German aircraft developments, she was told she could be of immense value to the United States, where she would become rich, even more famous, and could fly anything she liked. Otherwise, she would be imprisoned and brought to trial. She refused the invitation: she would rather die in prison, unable to fly, than betray her country by flying without honour as a free woman. For fifteen months, she was a prisoner of the Americans in Germany, indignantly surprised at the anti-German attitude of her captors. It was at first suspected in the absence of a body to prove that Hitler was dead that she might have flown him out of Berlin.

Hanna Reitsch's conviction that there was a deliberate anti-German smear campaign included a claim that her own part in the last days of the war had been falsified. According to her account, she was pressured, while a prisoner, into attending a press conference at which, as she had never been a Nazi party member, she was expected to denounce Hitler and his régime. Proudly wearing her two Iron Crosses she refused to speak against her country or her leader, in spite of her fear of serious repercussions if she did not co-operate. An account of her time in the Bunker was subsequently published both in the press and in the book *The Last Days of Hitler* by the historian Hugh Trevor-Roper, an intelligence officer in Berlin at the end of the war, and perpetuated by other writers who used his book as their source. Hanna Reitsch denied all knowledge of the supposed eyewitness account of Hitler's last days included in Trevor-Roper's book, although it was written in the first person and bore her name. She was further incensed, a quarter of a century later, when the film *Hitler: The Last Ten Days*, starring Alec Guinness, came out in 1973 based on the version to which she objected. Once again, she complained, she had not been contacted for her story. She had, she said, never talked to Hitler and Eva Braun together, there had been no dancing and drinking in the Bunker, where the atmosphere had been 'deeply depressing . . . quiet as a grave', and her relationships with Hitler and von Greim had been strictly professional.

In the first few years after her release in November 1946 Hanna felt constantly persecuted. Rumours were, she claimed, deliberately spread about her supposed loose living: people she had never met said that she had had sexual dealings with various leading Nazis, including Hitler, von Greim and Goebbels, and invented stories about how they and their friends had slept with her. She was supposed to have taken payment in kind, living a life of luxury and driving a large American car. Certain that these were the repercussions of her refusal to co-operate with her American captors, she was frequently tempted to take her own life. Newspaper articles appeared purporting to be based on interviews which she said had never taken place. A letter of support, written and signed by some of her friends, was sent to newspapers but never printed. When a journalist who knew her brother interviewed her at length and wrote a series of articles which were to present what she had persuaded him was the truth, the editor refused to publish anything which might be construed as 'a glorification of the Third Reich'.

Even after she had withdrawn to lick her wounds and write the first part of her autobiography, Hanna continued to feel persecuted. She claimed that the book *Fliegen mein Leben (Flying my Life)*, which was published in Stuttgart in 1951, was removed from the shelves of major bookshops because of anonymous threatening letters. Living largely on the charity of friends, with a small income from sales of the book, to give her life some meaning she became involved in Samaritan-type counselling. She did not at first even have the solace of flying, as Germans were not allowed to fly in the years after the war, but managed to renew her licence to fly gliders and small powered aircraft in Switzerland.

In 1951, there was an emotional reunion of German pilots, and permission was given for glider clubs to start up again. The following year, Hanna was a member of the German team which competed in the World Gliding Championships held in Spain, and in the next few years she competed in gliding competitions in Germany and broke several more records. In 1958 she was, however, refused a visa to compete with a German team in Poland: the rest of the team competed without her. Incensed at the lack of 'honour' shown by her teammates, Hanna quarrelled with the German Aero Club, and for the next twenty years refused to participate in any German gliding activities although she enjoyed her international status as a glider pilot in many other countries.

In India, she was Nehru's personal guest and took him gliding; she was invited to America and received at the White House by President Kennedy. She started and, for nearly four years ran, a gliding school in Ghana, became an honorary member of the Californian Society of Experimental Test Pilots, joined the international association of women helicopter pilots known as the Whirly-Girls, and in 1972 was elected Pilot of the Year in Arizona. In 1978, when she was sixty-six, she at last settled her differences with the German Aero Club after a ten-hour flight of 715 km which set a women's world record, and continued flying until a few months before her death in August 1979.

Hanna Reitsch is now considered by many other pilots to have been 'the greatest' for her undoubted courage and skill. She was also capable of great generosity to aspiring pilots of a younger generation: Sheila Scott, for instance, the British record-breaking pilot of the 1960s, remembers with gratitude the encouragement she was given by Hitler's former test pilot. Politically, Hanna Reitsch remains controversial: did her fanatical devotion to her country blind her to the more inhuman activities of the Nazis, or was she willingly involved with her leaders beyond the call of her duty as a test pilot? Opinions differ, but there is no doubt about the intensity of her patriotism.

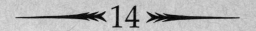

14

Jacqueline Cochran:
Rags to Riches and Records

Jacqueline Cochran was relieved when she discovered that she was an orphan and was only the adopted daughter of the improvident and slovenly couple she knew as parents but who could hardly be said to have brought her up. The exact date of her birth was a mystery: she gave it variously as 1906 and 1910, and presumably did not know it. Later, revelling in wealth and prestige, she often behaved outrageously; her behaviour was condoned by those who admired her for the tenacity with which she had struggled from the bottom to the top of American society.

As a child, she was dressed in old sacking, slept on the floor, was allowed to run wild, barefoot and without schooling, and supplemented a meagre diet by searching for crabs and clams. She wandered from one poverty-stricken Southern sawmill town to another with her adoptive family, and went to work at an early age in a sawmill: she was probably eight, and was paid six cents an hour for a twelve-hour night shift. It was not long before she left home to better herself, working at a variety of jobs in hairdressing, beauty and cosmetics until, at the approximate age of fifteen, she was earning more on commission as a permanent-wave operator than her boss. Her ambition was simply to earn enough for clothes, a car, and travel.

She had the car and clothes when, encouraged by one of her customers who felt that she was capable of better things, she did a three-year nurse's training and then took a job with a doctor in a sawmill town. The sickness, misery and poverty she encountered were a reminder of her childhood, and persuaded her that she could do nothing to help anyone, herself included, without money. She returned to the hair and beauty business, working her way up until she moved to New York, which seemed the best place to make her fortune. Charles of the Ritz was prepared to employ her only if she cut her long curly blonde hair, but Antoine, whose Saks-Fifth Avenue Salon was the height of fashion, took her on complete with hair.

Soon, working for Antoine's establishments both in New York and Miami, and often driving virtually non-stop between them, she was earning good money and had been taken up by a high-powered social set. At a party in Miami in 1932, she met Floyd Odlum: as soon as she saw the 'thin, clean-cut man', she felt that she had met her destiny. Quiet and serious, the son of a Methodist minister from Ohio and many years her senior, he had moved rapidly from his first job, selling shoes, to make several million dollars by the age of thirty-six as a Wall Street lawyer and investor. At their first meeting she told him of her plans to 'go on the road' for a cosmetic company: she took his comment seriously that to keep ahead of the competition she would 'almost need wings'.

That summer, she used her holiday to learn to fly. When she paid for her first lesson from her hard-earned savings, 'a beauty operator ceased to exist and an aviator was born', although the two co-existed successfully for the rest of her life. In three weeks she had her licence, taking the usual written examination orally as she was still barely literate. She set out almost immediately to fly to Canada in a rented Fairchild, to the alarm of flying acquaint-ances who realised better than she did how inexperienced she was. She had never seen a compass and had little idea of how to map-read, and when she landed to ask the way to Montreal was given some instant instruction about how to follow a compass course. When a naval friend, Ted Marshall, offered to teach her to fly 'the Navy way' if she had her own aircraft, she took up the challenge and bought an old Travelair with a Gipsy engine for $1200.

Armed with her commercial licence, in 1933 Jacqueline started a beauty salon in Chicago with an allied laboratory in New Jersey, flying frequently between the two and making a satisfactory profit. She sold her Travelair and bought a new Waco, intending to enter the 1934 England-Australia race with Ted Marshall. When he was killed she flew instead to his funeral in the Waco with her four-year-old niece, daughter of her 'sister' Willie Mae, crashing on the way: neither she nor the child was injured. In the autumn of 1933, she flew Floyd Odlum to Florida, an alarming flight in murky weather which persuaded her of the need for some intensive instrument training. When the Air Force took over airmail contracts, she employed an ex-airmail pilot to teach her.

She bought a Gamma, again for the Australia race, adding extra fuel-tanks and a prototype supercharger which was supposed to increase its speed from 220 mph to 240 mph: it failed over the Arizona desert, nearly asphyxiating pilot and co-pilot, and when it let her down again over New Mexico she crash-landed. It was with a new, modified Gamma which was neither finished nor tested that she eventually started the Australia race, sitting on a crate as the pilot's seat was not ready. The idea was that she would prove its reliability during the race and would receive royalties on any aircraft later sold by its manufacturers, Granville Brothers, who were hoping for a military contract.

Before the race, however, she had to undergo the first of seven major stomach operations: she flew herself to the hospital, which she left without medical permission after eight days. The Gamma proved 'squatty, fast and most unstable', and after a forced landing in Bucharest with engine trouble Jacqueline and her co-pilot had to abandon the race.

Jacqueline and Floyd were married in 1936. Before the marriage, she hired a private detective to trace her natural parents, and handed his report, unread and in a sealed envelope, to Floyd: neither of them ever opened it. Floyd's money, business acumen and contacts, added to Jackie's independent and growing wealth and her flamboyant and energetic determination, helped her to have her own way: she hated to be thwarted. With his encouragement, she raced and broke records, often trying some new device or development in aircraft which had not been previously fully tested. An extra glow of pride was added whenever her experience had a useful outcome, as when in 1937, in a fabric-covered biplane with neither heating, pressurisation nor oxygen mask, she burst a blood vessel in her sinus: part of the cumulative evidence which led later to cabin pressurisation and the use of oxygen masks.

Among her many successes, she was the first woman to win the Bendix race, in 1938, in a prototype Sikorsky pursuit aircraft which she had to fly with one wing higher than the other to ensure an adequate petrol supply: a wad of paper was blocking the fuel pipe. One of the judges drove out to meet her on the runway, but had to wait while she attended to her make-up. The instincts of a beautician never deserted her, even after the most strenuous flight, and some of her most successful beauty products were developed to combat dry skin and chapped lips in the air.

The Odlums' base was a ranch in Southern California, where before their marriage they had each, independently, bought land: Floyd 1000 acres, and Jacqueline a mere 25 acres, combined as a working ranch complete with private golf course, stables and an Olympic-size swimming pool. Entering the Coachella Ranch after a three-hour drive across arid desert was 'like driving into the Garden of Eden', with heavily scented tangerine and grapefruit trees, rich green grass and exotic shrubs. An army of servants waited on the Odlums and their guests, among whom were presidents, generals, colonels, film stars, dukes, duchesses: anyone who was influential and interesting might be invited to stay. Jackie, dressed in the latest Paris fashions, jewels and furs, treated them all equally disrespectfully, and expected them to fall in with her wishes, whether to play golf, to be driven by her at breakneck speed, or to drink and gamble after dinner. A visit to the Coachella Ranch could be as exhausting as it was exhilarating, and invitations were sometimes refused by people who found Jacqueline Cochran overpowering.

Floyd and Jacqueline each had a secretary and personal maid, but shared a private switchboard and telephone operator. A separate line was installed for calls to Howard Hughes, a personal friend and business associate who could not bear the possibility of being overheard by the operator. Jackie often spent the morning in her room telephoning friends all over the world. Her life-style was as far removed from the poverty and squalor of her childhood as it could be, and she developed a fetish about cleanliness, taking several showers a day and insisting on daily clean sheets.

It was at the Coachella Ranch that Amelia Earhart and her husband relaxed before Amelia's fatal round-the-world flight in 1937. Jacqueline Cochran tried to dissuade her friend from setting out: her hunches often turned out to be disconcertingly accurate forecasts. Indeed, Floyd Odlum, who had made a study of psychic phenomena, was convinced that his wife possessed extrasensory ability. They claimed to be able to communicate with each other from a distance and in their sleep, and on several occasions Jacqueline was apparently able to say correctly what Floyd had been doing when they were apart. When George Putnam enlisted the help of Jacqueline's psychic power in the search for his wife's aircraft in the Pacific, for several days she gave an account of what was happening to Amelia and her co-pilot, the accuracy of which, as they were never found, could not be established. Jacqueline, a devout Catholic with a firm belief in an after-life, lit a candle for Amelia when her senses told her that she was no longer alive, and was so shaken by the experience that afterwards she rarely tried to use her powers.

At the beginning of the Second World War, civil pilots, including Jacqueline Cochran, were grounded: she shared her time between her successful cosmetic business and the ranch. She was determined, however, to find some way into the war effort, and in 1941 persuaded friends at the top – General Arnold in America and Lord Beaverbrook in England – to let her fly a bomber across the Atlantic. Although England certainly needed the aircraft, the military pilots were so incensed at the idea of a civilian, and a woman at that, being allowed to undertake such a flight that she had to let one of them take off and land. In England, she called on Lord Beaverbrook, Minister of Aircraft Production, and met Pauline Gower, leader of the women ferry pilots in the Air Transport Auxiliary.

When she returned to America Jacqueline was invited to the White House to discuss the possible role of women pilots in the war, first with Eleanor Roosevelt and then with General Arnold and General Olds, who was already forming the American Ferry Command, the forerunner of Air Transport Command. It was, the men decided, not yet time to use American women pilots at home, but two weeks later Jacqueline was invited to take a contingent to Britain to join the ATA. Of approximately 700 licensed women pilots in America, there were about 75 with the necessary experience. Jacqueline Cochran sent a 2ft-long telegram to forty of these, and was given the honorary British title of Flight-Captain when she arrived in London in time to greet the first five American women to join the ATA in April 1942.

Her flamboyant wealth and habit of going to the top to

circumvent regulations made her unpopular with the established British ATA women, particularly as she had been given a rank which they considered undeserved. Many remember her chiefly for turning up at an airfield in a Rolls-Royce wearing her second best mink coat, a tactless display of wealth in a country which had suffered the hardships of war for two-and-a-half years, and where petrol and clothes, as well as food, were rationed. Although based at White Waltham, Jacqueline stayed at the Savoy Hotel, and had a London flat where the American ATA pilots, to whom she was motherly and protective, could meet and relax when they were not on duty.

She stayed in England only until she thought the time was right for women pilots to be used in her own country, but when she returned to America in the autumn of 1942 she found that Nancy Love, a rival whom she nevertheless called 'an exceedingly fine pilot', was already organising a Women's Auxiliary Ferrying Squadron as part of the Army Air Force's Transport Command. After some wrangling, it was agreed that, while Nancy Love continued her operation, Jacqueline would organise a training scheme for women with less than the 500 hours' flying originally required. She insisted on a strict military régime, although even after the two groups merged as Women Airforce Service Pilots in August 1943, they remained technically civilians. The WASPS, with Jacqueline Cochran as Director of Women Pilots and Nancy Love as their executive on the staff of Air Transport Command's ferrying division, were not only ferry pilots: they also towed targets for student anti-aircraft gunners, took part in smoke-laying during exercises, engineering test flights, simulated gas attacks, day and night training missions for radar and searchlight operators, and simulated low-level strafing or 'legalised buzzing', which was particularly enjoyed.

Jacqueline Cochran spent only a few days at her home and took little time off even for two more major operations. When the WASPS were disbanded in December 1944, her report claimed that women had proved conclusively that they were as fit psychologically and physiologically as men to be pilots. She kept in touch with many of the wartime women pilots, and ten years later estimated that a quarter of them were still earning a living in aviation.

At the end of the war, Jacqueline used her influence with

high-placed friends to go as a reporter to the Pacific. In Manila she was lent a flying boat so that she could fly over the islands, and although women reporters were banned from the surrender ceremonies in the Philippines she was allowed in as a service woman. As hostess to Archbishop Spellman, she met the first batch of repatriated prisoners and could hardly believe human beings could be in the condition she saw. She was the first American woman to land in Japan after the war, and tried three times to visit Hiroshima, but was foiled on each occasion by bad weather. In Tokyo, where she found 'destruction and desolation', she made a survey for General Arnold of the role of Japanese women in the war: although she found no evidence that they had played any part, she discovered a file on herself and several on Amelia Earhart, who had been rumoured to have survived her Pacific crash as a Japanese prisoner.

Her next stop was Shanghai, with a flight crew which included three colonels, three lieutenants, three majors and two captains. Madame Chiang Kai-shek presented her with Chinese Air Wings in recognition of her war work, and through Floyd's cousin, a general, she talked for two hours with Mao Zedong. In Peking, a Japanese general gave her his Samurai sword before his formal surrender. She then visited India, where she flew round the Taj Mahal by moonlight, and Egypt, Palestine and Persia – where she met the Shah, 'gentle, shy and very attractive'. In Rome, she was granted a 28-minute audience with the Pope, the memory of which she cherished more than any meeting with royalty or Heads of State.

In November 1945, Jacqueline reached Germany: through the US Supreme Court Justice Robert Jackson, she was admitted to the courtroom in Nüremberg at the start of the Nazi war trials where she gloated patriotically over 'Hitler's cohorts' in the prisoners' dock. At a press conference Jacqueline asked the German defence lawyer if he thought his clients were getting a fair trial, which he answered was impossible with no German on the tribunal. To her question about his personal involvement with the Nazi party, he replied that he had not only been a member, but a prominent one: the courtroom had to be cleared because of the ensuing outraged uproar.

Jacqueline's tour continued with a visit to Buchenwald on the way to Berlin, where 'misery crowded in . . . from all sides'. She gave her food away as a gesture, and bribed her way with cigarettes into the Chancellery, which she described as a 'beautiful underground home' with air-conditioning. The story of Hitler's suicide she brushed aside as unlikely: 'even that takes courage, when it is clear that he had neither courage nor initiative left'. A landing strip in the Chancellery grounds persuaded her that Hitler had been flown out in a hidden aircraft. Three years later, during the Berlin airlift, Jacqueline flew into the city again with the crew of an aircraft carrying sacks of coal from Frankfurt.

An avid traveller, she was in Russia for two weeks just before the start of the Korean War, and found there 'fear, work and poverty', but 'nothing wrong with the Russian people as a whole that a higher standard of living and the truth won't cure'. In Spain, she and Floyd spent 45 minutes with Franco, whose 'rounded face and soft expression seemed better suited for smiles than harshness'. Floyd sold aircraft to the Spanish, and Jacqueline satisfied her motherly and do-gooding instincts by adopting three little Spanish boys whom she established in a home in Madrid, where she left her golf clubs as an extra incentive to visit them frequently.

Her travelling reinforced her belief in the American way of life. She decided she should serve American interests in Congress. When she was twice offered nomination as a Democratic congressional candidate, she turned down both invitations, and later attempted unsuccessfully to stand as a Republican. Instead she found a role on the political sidelines as one of General Eisenhower's most ardent campaigners. The 'I like Ike' slogan was coined at a meeting at her New York flat, and a vast rally which she and a group of like-minded friends organised in Madison Square Garden was instrumental in persuading Eisenhower, who was later a guest at the Coachella Ranch, to return from Paris because his country needed him.

Although the war had interrupted her record and competition flying, Jacqueline was among the competitors in the first postwar Bendix race, in 1946, in an ex-government Mustang. She always imposed a strict training régime on herself, with a 'strong meat' diet and arm and shoulder strengthening exercises, although she was naturally powerfully built with big-boned, strong hands. As the Bendix races had an early start, which meant being at the airfield well before dawn, for several weeks beforehand she went to bed earlier and earlier to acclimatise her body to a new sleep pattern. For at least a week before races, she practised instrument approaches and take-offs for an hour or two every day, hiring a Harvard training aircraft before the 1946 race in which, by flying it from the rear seat with severely restricted vision, she could simulate blind flying in the Mustang.

The race itself was fraught with problems. First, the radio failed. Near the Grand Canyon she tried to climb to 30,000 ft above bad weather, but at 27,000 ft the engine cut, then surged on and off alarmingly: it had been tested only to 25,000 ft, and above this height suffered from an irregular supply of fuel. The only answer was to dive into cloud, at speed and with the aircraft bucking so much that it was difficult to control.

After flying on instruments through the storm, Jacqueline decided to jettison the external fuel-tanks over the mountains rather than risk dropping them in poor visibility over a populated area. The drop mechanism had however not been tested, and although the front connection was released, the rear of the tanks stayed firmly attached, jerking the aircraft violently and swaying up to hit and damage the wings. Wondering whether she would reach Cleveland before the Mustang disintegrated, Jacqueline fought to keep control, but landed only six minutes behind the winner.

Other races and record attempts were equally fraught with danger. Sometimes, after calculating her fuel need to the last few gallons to avoid carrying unnecessary weight, she landed with enough left for only a minute or two in the air, and once, when she had to circle an airfield because a naval squadron was flying over it in formation, the engine cut just as her wheels touched the ground. On another occasion an aircraft broke in two as she landed: the two halves bounced and collided, but although she was covered with oil and petrol Jacqueline was able to board a commercial aircraft twenty minutes later so that she could meet Floyd for dinner in New York.

In 1947 Jacqueline met Captain Charles Yeager, the American air force pilot who had just broken the sound-barrier for the first time. The meeting was the start of a long friendship, although at first Yeager's wife Glennis resented the way Jacqueline took over their lives, and felt that her generosity was merely an attempt to buy her friendship as a way to influence her husband: Jacqueline invited them to stay at the ranch, gave Glennis her cast-off Paris clothes, and expected 'Chuck' to be at her beck and call. He, like many others, responded with exasperated and often exhausted affection and admiration. 'If I were a man,' Jackie told Chuck, 'I would've been a war ace like you. I'm a damned good pilot. All these generals would be pounding on my doors instead of the other way round. Being a woman, I need all the clout I can get.' Glennis felt that 'Jackie lived vicariously through Chuck, and Floyd lived vicariously through Jackie's exploits'. Floyd by then was crippled with arthritis, but was still Jackie's most devoted admirer, and always ready to use his clout on her behalf.

By the early 1950s Jacqueline's one unachieved flying ambition was to fly a jet. Although Floyd owned the company which built Sabres, it took her two years of negotiations to obtain permission, as a specially appointed consultant for Air Canada, to fly a Canadian Sabre, with which in May 1953 she was the first woman to fly faster than the speed of sound. Chuck Yeager trained her for the attempt, and even saved her life on a low-level flight when he noticed a fuel leak and was able to warn her by radio to land in time to avoid an explosion. The potential danger was reflected in the insurance cover of £10,000 for an hour's timed flying. In six hours' total flying time, Jacqueline made thirteen flights, broke three men's records, tried for a fourth, and three times flew faster than sound. The Sabre reeked of her perfume for months afterwards: she sprayed every aircraft she flew with scent to mask the usual cockpit and body smells.

Jacqueline's first supersonic dive, from 47,000 ft, was not recorded, so she did it again. Then, for the benefit of Paramount Films and *Life* magazine, she and Chuck Yeager, the fastest man and woman in the world, made simultaneous flights, climbing to nearly 50,000 ft before putting the two Sabres almost wing-tip to wing-tip into a full throttle vertical dive in a supersonic duet. Afterwards she was for once quiet, so thrilled that she was almost speechless.

Passing the sound-barrier was for Jacqueline a spiritual and emotional experience, although physically it also made its mark. She did not bother at first to wear a G-suit in the Sabre, and the gravitational force to which her body was subjected resulted in severe pain because of the many lesions left by her frequent abdominal operations. She was forced to go into hospital, but was back in the air in less than a week. As she flew through the sound-barrier, she felt confidence, humility and trust, a great alertness about what was happening to the aircraft and how to react to it, and a warm sense of accomplishment which she wanted immediately to share with Floyd.

Five years later, in 1959, she set new speed records in a Lockheed F-104 Starfighter, a Mach 2 aircraft which many fighter pilots found alarming. Later that year, she was elected president of the most prestigious international aviation organisation, the *Fédération Aéronautique Internationale*, and decided to fly to its annual meeting in the twin-engined Lockheed Lodestar which Floyd had given her. The venue was Moscow, and Jackie wanted Chuck Yeager to accompany her: as usual, she got her way. She travelled with a retinue of private secretary, maid, hairdresser, and personal interpreter.

In her fifties, Jacqueline continued to fly and to break records. On a flight in a Lockheed Jet Star from New Orleans to Hanover, she was the first woman to fly a jet across the Atlantic and established sixty-nine records. Wherever she went, she was treated like royalty, but it was not only for her own interests that she used her influence: she was capable of great generosity to her friends, and would stand no opposition in her efforts to advance their interests. On several occasions, she paid for and arranged medical treatment for the Yeager family, and when she decided that Chuck's flying achievements should be honoured she bullied generals and chiefs of staff until he was awarded a Medal of Honor.

There were few honours in aviation or business which were not at some time awarded to Jacqueline Cochran herself. In 1945, she was awarded the Distinguished Service Medal, normally only available to members of the armed services, and later rose from a lieutenant to a colonel in the Air Force Reserve. She had honorary wings from half a dozen air forces, and the French *Légion d'honneur*, as well as several honorary degrees and awards as Woman of the Year. The cosmetic company which she had started from her hard-earned savings developed into a business with annual sales of several million dollars.

A dozen major operations – seven on her stomach, one on her sinuses, and three on her eyes – failed to curb her energy: after one, she even recuperated by dancing at the New York Roxy

Picture Theatre. In her sixties, a serious heart condition and a pacemaker finally forced her to give up competitive flying and sell her Lodestar. Although Floyd had patiently put up with his frailty and ill-health for many years, Jacqueline was far from tolerant of her own weakness. After Floyd's death at the age of eighty in 1977, she declined rapidly, swollen by heart and kidney disease which forced her to sleep sitting up in a chair. She became 'so impossible' that few of her friends apart from the Yeagers visited her in the modest house to which she moved when the ranch was sold to a property developer.

Jacqueline Cochran died in 1980. Only fourteen people attended her funeral, a poor tribute to the fastest woman in the world, friend of four presidents of the United States, honoured guest of royalty, who from her barefoot childhood had achieved successes both in aviation and in business which would have been remarkable whatever her origins. As Chuck Yeager put it, 'when Jackie Cochran put her mind to do something, she was a damned Sherman tank at full speed'. Others found her dynamic and flamboyant, warm-hearted but domineering. Her husband called her simply 'the most interesting person I have ever met'.

15

Jacqueline Auriol:
the President's Daughter-in-law

For twelve years, from 1951 until 1963, the daughter-in-law of the president of France competed with Jacqueline Cochran for the title of the 'fastest woman in the world'. Jacqueline Auriol, who started flying as an escape from constant publicity and formal life at the Elysée Palace, was severely injured in a crash after which the will to fly became for her synonymous with the will to live, and became the only woman test pilot in the world.

Jacqueline was studying art in Paris when she met Paul Auriol, a Political Science student whose father's militant socialism and atheism were the antithesis of her own family's staunch monarchist and Catholic beliefs. Both families were equally opposed to their engagement, and it was only the imminence of war that persuaded them to allow the two twenty-year-olds to marry in February 1938.

Jacqueline's first baby was a few months old when Germany invaded Norway and Denmark; throughout the war the Auriols' family life was disrupted by danger and hardship. Paul's father was imprisoned for eight months for his open opposition to Marshal Pétain and the Vichy government, and was allowed back to his home only under strict supervision. The whole family nevertheless became deeply involved with the French Resistance movement, and Jacqueline fled frequently across France with her two young sons while she and her husband adopted various names. When the war ended, their two sons were told for the first time that their surname was Auriol.

Vincent Auriol became President for Foreign Affairs in the post-war advisory assembly, then swiftly Minister of State, President of the Constituent Assembly and President of the National Assembly. The luxury of the family's new life after the years of secrecy and fear had an almost unreal quality, culminating in January 1947 with Vincent's election as President of France. Jacqueline entered wholeheartedly into her role as stand-in president's hostess for her mother-in-law, welcoming the chance to mix with royalty, international political, intellectual and artistic leaders, to dress up and be admired. She and Paul were swept into a social round, and like 'a couple of big kids with a new toy' were ready to dress up in black tie and tails and new couturier dresses every evening. Then their initial popularity was transformed into notoriety: in a smear campaign to damage the reputation of the president, rumours were floated about the source of his son's and daughter-in-law's apparent wealth. Their names were linked with black market rackets, starting with cars, progressing to gold, diamonds, penicillin, and finally drugs. When Paul and Jacqueline started flying in 1947, as an escape from notoriety and the atmosphere of the Elysée, it was claimed that Jacqueline was using a private aircraft to smuggle drugs.

She was initially unenthusiastic about flying and carried on only to please Paul. As the technicalities became less incomprehensible her interest grew, until after five or six hours in the air, as she put it, 'It clicked. When I adjusted the controls, my movements came simply and naturally. The pilot pupil found she was in tune with the machine in its own element.' She decided that she could master what she still considered no more than a sport, and in March 1948 received her French Private Pilots' Licence – the *brevet du premier degré* – then immediately started working for her *brevet du second degré*, which she gained after 41 hours' flying at the end of the following month. She used her status to gain access to military aircraft at GLAM, but to fly there seriously needed to pass various military pilots' tests, including one in aerobatic flying.

Raymond Guillaume, one of France's leading aerobatic pilots, was persuaded to take her on as a pupil only after putting her through a gruelling initiation to make sure that she was not merely a society woman satisfying a passing whim. To equal Guillaume's supreme ease and mastery in the air became Jacqueline's ambition. Flying had become an obsession with her, an end in itself, and not merely a means of escape from her public life, which was nevertheless occasionally an asset. When she was invited by the Air Minister to give an aerobatic display at the town of which he was mayor, it was because the president's flying daughter-in-law would attract attention to the event. The display, after two months' intensive preparation under Guillaume's demanding guidance, was a success, and Jacqueline Auriol felt that she had at last accomplished something through her own hard work.

A week later, she left her husband and children on holiday in Brittany to accept another invitation, for a publicity flight as a passenger with Guillaume and a pilot from the *Société de Constructions Aéronavales* in a prototype amphibian, the Scan 30. All went well at first, then the pilot inexplicably crashed the aircraft into the river: a mirage, an error in judgement or a mistaken impulse to show off could all possibly have been to blame. The pilot and Guillaume escaped with relatively minor injuries, but Jacqueline's head took the full impact of the crash. She felt no pain as she floundered in the water and was pulled out by press photographers, and realised only that the injuries to her face would leave her permanently disfigured. Her first thought was cosmetic surgery, and she asked by name for a well-known surgeon. With no idea that her life was in danger, she next wanted to know how soon she would be able to fly again: flying, which had just hurt her, was to make amends.

The president was summoned to her bedside, and he in turn summoned specialist after specialist. There was little that the cosmetic surgeon could do for the pulp which had been the face of 'the prettiest woman in Paris' and which was deformed by countless fractures. In spite of the attention of so many experts, the bones did not knit, and Jacqueline refused to allow her sons, aged eleven and nine, to see her: the family was separated, except for one brief visit while she was still swathed in bandages, for two years. Even when her physical strength began to return, there was still no improvement to her face, and she lived in a lethargic despair relieved only when Raymond Guillaume took her for a secret flight in a borrowed Fairchild 280. The discovery that she could still fly gave her new hope.

When the doctors suggested that the reason she was not getting better was because she did not want to, she wondered if they were right. To discover, she went alone to a mountain retreat, taking textbooks on mathematics, aerodynamics, anatomy, a human skull – she was determined to find out all she could about her skull and its refusal to heal – and a loaded revolver. She was prepared to shoot herself if she discovered that she did not have the will to live, but in her solitude it did not take her long to decide that the talk about her subconscious desire not to recover was 'a lot of nonsense'.

Yet another doctor was summoned, an army surgeon who specialised in rebuilding *gueules cassées*. For three months he undertook a series of painful but successful operations. Jacqueline was still disfigured and hated to be seen, but was able to start flying again, with the unprecedented privilege of flying as co-pilot on various aircraft at the French Flight Test Centre. In the air, her depression vanished and there was no need for self-consciousness about her appearance. On the ground, she started on a series of sixteen operations in America with a specialist in 'repair surgery' who, over a period of six months, gradually gave her a new cheekbone and a new nose. After several more visits to America, and yet more operations, she at last had a recognisable face again: not quite as it had been before, but good enough to allow her to resume life with her family.

During her progress from clinic to clinic, hospital to hospital, operation to operation, Jacqueline had formulated an ambition: to have a career in aviation, as a test pilot at the Flight Test Centre in Brétigny. First she needed to attract attention to her ability and dedication by setting a record: the Chief of Staff of the Air Force, a personal friend, agreed to put France's only jet aircraft, a British Vampire fighter, at her disposal for an attempt to break Jacqueline Cochran's women's world speed record. Guillaume once again supervised her training, first in a Morane two-seater practice jet and then in the single-seater Vampire, both far faster than anything she had flown before.

After fourteen flights in the Vampire, Jacqueline was ready to challenge the four-year-old record of 441 mph. Flying at high speed with a record at stake was more than an individual effort, requiring precision and total concentration from a team of which the pilot was only one member. On 11 May 1951, she averaged 69 mph faster over the 63-mile course than the American Jacqueline, who proposed her for the Harmon Trophy: it was presented to her for the first time – she later won it again twice – during one of her visits to America for operations on her face.

The ability to work as part of a team and to sustain total

concentration was essential for a test pilot. Jacqueline had proved that she could do both, and her record, the first to be won by a French pilot since the war, enabled her to gain her military licence and be accepted as a *pilote de servitude* at the Flight Test Centre. She was still, however, a long way from becoming a fully fledged test pilot: *servitude* consisted of liaison flying between towns and acting as practice target for radar control or simulated aerial attack. She still had to become a glider pilot, obtain instructor's licences for gliding and powered flight, and work at instrument flying, before she could even apply for a place on a training course.

When she had all the necessary preliminary qualifications, General Bonte, the head of the Flight Test Centre, grudgingly allowed Jacqueline to fly at Brétigny as a 'pilot seconded from light aviation'. Test pilots, the flying élite, were traditionally male, and the General could not admit even the President's daughter-in-law to an organisation run almost as a club until both he and her potential colleagues were certain that she would fit in. Jacqueline was thirty-five when she was allowed to go back to school for eight months' intensive preparation for the entry examination, followed by another eight months to prepare for the final examination. She redecorated her bedroom in black and white so that there would be no distractions while she studied, and at last became test pilot no. 29.

As one of the official Flying Personnel at Brétigny, she had achieved her ambition of working on equal terms with men. She neither wanted nor received any concessions, although she was asked to decorate the pilots' lounge. As soon as she entered the office she shared with Guillaume, she became a new person, ready to apply herself to whatever task awaited her. Changing into her flying suit, she reported first at the met. office for a detailed daily weather report, then collected her parachute, which she always insisted on carrying herself, the day's flying orders and a test programme sheet. The only tests not routinely allocated to Jacqueline were those on such military details as machine guns, guns and rockets. Her favourite tasks involved modern ultra-fast aircraft, and only a few months after Jacqueline Cochrane broke the sound-barrier, Jacqueline Auriol became the second woman to do so, on a half-hour test flight with a Mystère II. When the needle on her machometer passed Mach 1, the invisible barrier, her earphones were filled with shouts of joy from the ground, where her colleagues had heard the sonic boom several miles away although in the cockpit she had heard nothing.

Even more exciting was her first flight with the Mystère III, with which she reached Mach 2, twice the speed of sound. She had waited for a week, learning everything possible about the aircraft and sitting in the monitoring room while other pilots were flying it. The unaccustomed steepness of its climb was a revelation, and in three minutes Jacqueline was at 40,000 ft, flying so fast that she could send back information only every 10,000 ft instead of the usual every 5000 ft. She was hardly aware of crossing the Mach 1 sound-barrier. As she reached Mach 2, nearly 1500 mph, she could not restrain a shout of excitement, and on her final approach afterwards did a series of victory rolls over the airfield.

In October 1956 Jacqueline flew a Mystère IV for the first time. She was to demonstrate it in Germany the following day, and had been warned by a colleague already familiar with its peculiarities that it had an 'unhealthy' spin in which, because of its swept-back wings, it behaved differently from other aircraft. She broke the sound-barrier successfully twice, but on the third attempt the aircraft went into a tailspin, her oxygen mask became disconnected, and for a few seconds she blacked out. As she came round, she was convinced that she was approaching death, and discovered to her surprise that she had not lost her religious belief: she was not frightened, but merely curious about what would happen when she died. Then, still only half-conscious, her instincts took over: she remembered her colleague's instruction about how to get the Mystère out of a stall, plugged in her oxygen supply, and regained control of the aircraft just in time to avoid the crash anticipated by everyone on the ground. In spite of her intimations of immortality, she felt considerable relief at being alive, and apprehension about the following day's public performance in Germany: she nevertheless managed to break the sound-barrier and perform high-speed aerobatics as required.

Although the task of a test pilot was to eliminate danger, there were sometimes accidents, occasionally fatal. Jacqueline witnessed the death of a colleague while she was waiting to give an aerobatic display at a large air show at Cambrai. She, Guillaume and an expert on ejector seats were watching a Mystère II while the crowd waited for it to break the sound-barrier. The pilot was at 42,000 ft when his voice came over the loudspeakers saying calmly that he had 'lost control of the plane' at almost the same moment as the sonic boom was heard. With no audible sign of emotion, he told his colleagues on the ground first that the ailerons had 'gone crazy', then, when Guillaume instructed him to bale out, that the release mechanism of the canopy was jammed. For three or four minutes, he tried everything the three on the ground could suggest to release it, but without success, while the Mystère reeled drunkenly at over 600 mph before crashing on to houses in a suburb of Valenciennes. Miraculously no one else was either killed or injured.

As usual after the death of a colleague, the other nineteen pilots at Brétigny mourned him by not flying for two days, and his funeral was attended by test pilots from other centres: it was only at funerals that the country's test pilots, both past and present, were ever gathered together in one place.

Between 1951 and 1963, Jacqueline Auriol won the title of 'fastest woman in the world' five times; each time, Jacqueline Cochran won it back. The French pilot's 1963 speed was 1266 mph in a Mirage IIIR: well over twice the speed she had reached in the Vampire when she first took the record from her American rival. Each attempt involved months of preparation, starting with the sometimes lengthy task of obtaining official permission before embarking on the preliminary teamwork essential for success. The incentive was a mixture of 'because it is there', and pride in both a national and a personal achievement. In the air in a high-speed jet, wearing an uncomfortable but essential G-suit, Jacqueline had to concentrate at the same time on innumerable details, both inside and outside the cockpit: the many instrument dials on the control panel, the weather conditions, the voice telling her through her earphones when to alter course, by how much, how wide or tight to turn. Her speed records were made over a pear-shaped closed circuit course and closely monitored by time-keepers, engineers, mechanics and colleagues in a control room equipped with a computer: Jacqueline could detect from the tone of their instructions when they could see success on their screen before she could be sure of it in the cockpit. The flight itself took no more than minutes, during which the mental effort was so great that afterwards she was more exhausted than after any long flight, and the intoxication she felt when she succeeded was the result of months of dedication.

She was forty-five when she made her last major record flight, which, as usual, was beaten by Jacqueline Cochran. Although Jacqueline Auriol acquired two women's speed records for executive jets in 1967, and continued to do her job with unabated enjoyment, she subsequently devoted much of her time to demonstrations of aircraft abroad, as an aerial ambassador for France, for flying, and for women's role in aviation.

Her fame as a woman pilot was both better deserved and more welcome than her former fame and notoriety as the president's daughter-in-law. She had earned her place in aviation not through her social position, although at times that had certainly helped, but on merit, as a pilot accepted and respected by her colleagues. Theirs was the recognition she had sought when flying, after nearly destroying her, became her life.

≪16≫

Sheila Scott:
on Top of the World?

At a party in 1959 Sheila Scott announced to her friends that she was going to learn to fly. It was said on the spur of the moment, because she was bored, and was greeted derisively. She had, after all, failed her driving test twice. For years, she had been searching for something to give her life a meaning, alternating between elation and despair: little did she expect that she would find it in the air.

Seeing herself as a modern Florence Nightingale, she had tried nursing at a military hospital during the war. As a relief from daily discipline and distress, she next tried acting, as a stand-in for Deborah Kerr at the Denham film studios. Convinced that she too would become a star, but lacking the necessary dedication, she went to drama school, had several small parts in films, and worked spasmodically in repertory at Windsor, Watford and Aldershot, filling in between engagements by writing, modelling and designing clothes. She drifted into marriage with Rupert Bellamy, and took six years to extricate herself when that too failed. For a while, Buddhism, a Bohemian life-style and a new lover seemed to be the answer, until her guru disillusioned her by making advances.

Although the memory of a short flight with Alan Cobham's air circus as a sixth birthday treat was among the happiest of her unsettled childhood, by the time she reported to Elstree for her first flying lesson she was regretting her bravado. Her tight skirt made climbing into the cockpit difficult, she ruined her best nylons, and wondered what was wrong with the aircraft that it needed so much checking. She was too tense to enjoy the flight, but by the time she landed had decided to carry on. Her first solo was a success, but by the next morning she had both arms in plaster after slipping while she was clearing up after a party.

As soon as she had her private pilot's licence, Sheila bought a Jackaroo, an old rebuilt Tiger Moth. To the amusement of the male pilots, she insisted on having it painted blue with silver wings and white upholstery: she wanted her aircraft to be as feminine as herself, and christened it *Myth*, Greek for a female moth. No longer bored, she flew to North Africa alone, stopping every two hours to refuel, competed in her first race and won her first trophy, then spent the summer flying round Europe to races and rallies and the winter as the only woman on a night-flying course. By the end of 1960, she had accumulated 300 flying hours and passed all the tests for her commercial licence except her medical.

The board decided that she was too short-sighted to fly commercially, although she was allowed to demonstrate Cessnas for a dealer and retook her medical in America, where her sight was considered adequate. For a while she was commercially qualified there but not in Britain; this meant that she could fly into England as a paid commercial pilot, but could only fly out, unpaid, as a private pilot. By the time she eventually passed the British eye test through a new examiner, she had decided that record-breaking rather than commercial flying was to be her career, sold *Myth* to a parachute group and, in a borrowed single-engined Piper Comanche, temporarily called *Myth Sunpip*, broke fifteen European light aircraft records in 36 hours.

Only two women had ever flown round the world, Jerrie Mock and Joan Merriam Smith, both American. Jerrie had taken a northern route, and Joan had followed the longer route attempted by Amelia Earhart. In 1966, Sheila decided it was time for a European to take up the challenge, and to make it an even greater one chose the longest possible route, the 32,000 miles round the Equator, never before attempted by a solo pilot. At first, she planned to use *Myth Sunpip* and, as the Piper company appreciated the publicity a successful record venture could bring, was able to acquire the £18,000 aircraft for a deposit of £100. Then, because the fuel consumption of the Comanche 400 was considered inadequate for a Pacific crossing, *Myth Sunpip* was abandoned in favour of a lower horsepowered Piper Comanche 260, *Myth Too*, on equally favourable terms.

Sheila was advised about what to eat, what to wear – outsize clothes because on a long flight she would swell – and how to do exercises to prevent cramp in the confined space of the cockpit. She took a companion, a stuffed toy called Buck Rabbit, which also travelled with her on all her later flights. The solo expedition became an international team event, with mid-air encourage-ment from airliners over her radio and wide press coverage. Her reception at stopping places was warm and varied: in Damascus the cockpit was filled with scented jasmine; women in Karachi greeted her with garlands of flowers; she was given orchids in Singapore and necklaces of shells in Fiji; at the island of Canton, she was met by the entire population of fifty-one people. There were problems, including a burst fuel tank which drenched her with petrol over the Pacific, but remarkably few for such a long flight.

By the time she flew back into England, Sheila Scott was an international celebrity, besieged by reporters. *Myth Too* was displayed in Fleet Street, while she tried to recoup the costs of the flight. She felt more alone in the blaze of publicity – which meant that she had to employ three secretaries in her tiny London flat to deal with fan mail and invitations – than she had ever felt on her solitary 33-day flight round the world.

Soon after her first major flight, Sheila was offered sponsorship for another solo record attempt, to South Africa: it might, sug-gested Ken Wood, be equally advantageous to the pilot and to his new South African factory if he were to put up the money and she were to fly to Cape Town. She had only a few weeks to prepare herself for her departure. It was during the Arab-Israeli six-day war, and she had to turn back at Benghazi in Libya when a signature scrawled on her aircraft was thought to be a Star of David.

Avoiding the Middle East, she flew across the heat and turbu-lence of the Sahara to Nigeria, but could not land at Lagos because of an eruption of African violence. Flying on at night over mountains, she obeyed a sudden premonition that her height of 9000 ft was leading her into danger: when she landed she was told that, had she not done so, she would probably have flown into an unmarked mountain 9500 ft high. Like Jacqueline Cochran, she was convinced that she had some sort of sixth sense.

Her progress was broadcast over South African radio, and

when she reached Cape Town, beating the previous record by four hours, she was greeted by 10,000 people. Being a celebrity was disorientating: in the air, Sheila knew who she was, but on the ground she felt unable to live up to the superhuman image she had been given. She had many acquaintances, but few close friends, and yet another romance faded after three years during which she was constantly travelling to give lectures. When it came to a choice between a man and her aircraft, no man stood a chance, and she could not afford to turn down opportunities which would earn the aircraft's keep.

Fifty years after Alcock and Brown first flew the Atlantic, Sheila entered a race between London's Post Office Tower and the Empire State Building in New York, organised by the *Daily Mail*. The rules allowed competitors to choose any method or combination of methods of travel, starting from the top of one city's tower and finishing on the top of the other's. More than 300 people took part, several in both directions. Some competed seriously, some light-heartedly. Anne Alcock, the eighteen-year-old niece of the pioneer aviator, was first to leave from London, one of many, including Olympic sprinter Mary Rand, who travelled by commercial airline. Ben Garcia, a solo pilot, clowned his way into the headlines, even, although not intentionally, landing a Piper Cub upside down in a chicken coop, while the ageing Billy Butlin, founder of the famous holiday camps, travelled sedately in a morning suit and grey top hat. The BBC television cameraman Slim Hewitt set out with the aim of being last, and took more than 192 hours, travelling by sedan chair, bus, tube, coach and horses, train, hovercraft, balloon, various aircraft, motorbike, mule, bicycle, rowing boat, sleigh and covered wagon: he even tried hitch-hiking. Only one person gave up on the way.

Sheila was one of the more serious competitors, flying alone in *Myth Too* and adding four transatlantic records to her rapidly growing list. She started in the lift of the GPO Tower, leapt into a waiting Aston Martin with an unexpected escort of police motorcyclists, transferred to a Jet Ranger helicopter and then took off from Heathrow Airport for the Atlantic crossing. In New York, after battling with ice during the night, she reversed her helicopter, Aston Martin and lift routine. Although she had had no sleep for two days, she arrived looking patriotically immaculate in a red coat, white trousers and a blue hat, and won the £1000 first prize for a light aircraft piloted by a woman. Whitbread gave her a year's free flying: she had travelled with a horseshoe from one of the brewery's shire horses to bring her luck.

Few of Sheila's competition and record flights were as carefree as the Top of the Towers race. In the same year, on another flight to South Africa, she unwittingly became involved in what purported to be an attempt to free Nelson Mandela, but was later claimed to have been a plot by BOSS, South Africa's Bureau of State Security, to catch their prisoner in the act of escaping and so have an excuse to shoot him. Sheila, with no wish to become involved in any way in South Africa's politics, refused to have anything to do with the scheme, for which she was wanted only because she was famous enough to ensure publicity. Had the plan been carried out, with her participation, she would have been put on trial in an attempt to prove to the British the folly of their liberal attitudes.

Later in the year, on the 1969 England to Australia Air Race, things began to go wrong from the start. So many instruments failed that Sheila began to fear that it could not be coincidence: her autopilot, magsyn compass, transponder and fuel indicator were all out of action. The long-range radio was useless, with only half its trailing antenna, and when she overheard radio comments from other competitors like 'This race is apparently being run for the benefit of number 99', her number, she began to feel persecuted. In Singapore, where hers was the second aircraft to land, it was found that her aerial had been cut. When she was unable to transmit from either of her radio sets on the next leg, she again suspected sabotage.

When she left Singapore, her target of under three days and a new solo woman's record to Australia still seemed possible, but the weather could not have been worse: with neither radio nor autopilot, she was isolated in what soon became a struggle for survival. Her hands were raw from attempts to make her autopilot work, and it was almost impossible to fold or unfold her large charts single-handed in the confined space of the cockpit. She was uncomfortably hot and sticky in her lifejacket, her arms were aching with exhaustion and her legs with cramp, but the only way she could relieve the stiffness after sitting for over two days was by sitting on a small cushion, with no room even to stretch her legs.

Only a few hours from Darwin, lost, lonely, and frightened, searching desperately for a clearing in storm clouds, Sheila heard a woman's voice asking her to pray before she died. Then she began to pick up occasional Morse and conversation on the radio with the crippled aerial, but still could not make anyone hear her. Gradually she realised that the voices were talking about her, considering her chances of survival. Whether they were real or imagined, or a mixture of the two, was not clear.

At last, she saw land, and circled over a group of tiny islands searching for somewhere to put the aircraft down. Her troubles were, however, far from over: when she landed on a deserted airstrip on one of the larger islands, she was held up for several days by Indonesian bureaucracy and was unable to contact Australia or England, or to continue the race. She had had no sleep for nearly three days, was running a temperature, and was still unaware that the world's press was discussing her fate. Eventually, the Indonesians decided to be helpful rather than obstructive, patched up her radio, and passed on a message offering her an escort to Australia. She accepted gratefully.

Her escorts were three of the Red Arrow team, unofficial entrants in the race in a Marchetti, which although faster than Sheila's Comanche had a considerably shorter range. They planned to land in tandem at Kupang on the island of Timor so that the Marchetti could take on more fuel, but fog over both the coast and the mountains behind it forced them on through yet more bad weather. Finding it impossible to keep up with the leading aircraft, Sheila gave up the attempt and landed on Sumbawa, her relief to be alive less than her anxiety about the fate of her companions, who she knew would soon run out of fuel. Communicating with anyone either at Sumbawa or elsewhere proved at first an insoluble problem, but eventually, on Boxing Day, after an unhappy Christmas with dysentery to add to her misery, she was told that the men had reached Australia: they had landed on the only possible stretch of beach, where the local people had been able to provide them with enough fuel to complete their flight.

When Sheila reached Darwin, a week after she had hoped to arrive in triumph, she was accused among other things of having deliberately abandoned the men who had come to help her. The most serious of the stories which had been spreading during her disappearance was not the premature report of her death, but the unsubstantiated claim that she had attempted to obtain £10,000 in cash from the Singapore Royal Aero Club. She was increasingly convinced that there was a plot behind both the rumours and the mysterious damage to her equipment. At Alice Springs, she discovered that her manifold pressure pipe had been cut, and at Adelaide both she and *Myth Too* were hustled off the airfield, to the fury of the crowd waiting to greet her: it was said that she had asked for police protection. She was quoted at Griffith as having insulted the town, but denied the words attributed to her: 'Griffith is a bum town!'

Feeling ill and depressed as well as persecuted, she thought of suicide, but in Sydney, although she was accused of being a 'prima donna', she was otherwise treated gently by reporters, one of whom told her that her problems had been caused by jealousy among competitors, exacerbated by large amounts of money at stake for those who had put bets on the race. Her faith in human nature was to some extent restored on her leisurely return to England in *Myth Too*, on which all the necessary repairs were completed in America.

Having flown round the world both on the Equator and via Australia, Sheila dreamt of a third circumnavigation, across the North Pole. All her efforts to find a sponsor failed until an Adventure Trust was formed to finance the venture. As *Myth Too* was too small for the equipment she would need, Sheila ordered a twin-engined Piper Aztec, although she felt a sense of disloyalty to the Comanche which had served her so well. Even the Aztec would need intensive modifications, however, which were already under way when the sponsors decided to pull out. Rather than admit defeat, Sheila put *Myth Too* up as security against a bank loan.

The new aircraft was christened *Mythre*, and a moth was painted on its tail below the linked flags of Great Britain and the United States: Sheila was to undertake research for both countries. For the first time, a light aircraft would be tracked by satellite over the Arctic, levels of pollution in the earth's atmosphere would be automatically recorded, and her mental and physical state would be monitored through electrodes uncomfortably taped to her skin.

Flying in the Arctic presented considerable physical problems.

The fresh-air inlets to the aircraft had to be sealed against the intense cold, and Sheila needed five layers of clothing, topped by a mountaineer's survival suit and a sheepskin cover for the pilot's seat, as well as electrically heated gloves and boots to prevent frostbite. First, however, she had to fly through excessive heat. During the six months of intensive preparation, as usual a complex team effort, she had so little time to fly *Mythre* that the journey down to the Equator, where she had to turn round to start her expedition, was virtually a test flight. She tried to ignore the feeling that something was wrong before she left England. With *Myth Too* she felt at one with her aircraft, but she did not yet have the same close relationship with *Mythre*.

On the journey out, the engine persistently overheated, under-performed and used too much fuel. When disaster struck, while she was on the way up from the Equator, it was however nothing to do with the aircraft: in Malta, she was told that 'a few trophies' had been stolen from her flat. In London, where *Mythre* was to be winterised overnight, Sheila found that everything she possessed of either financial or sentimental value had been taken: jewellery, trophies, souvenirs, the videotape camera with which she had planned to make a film of her Polar flight, even log-books. It looked as if whoever had ransacked her flat had known it well. At least the equipment for the Arctic had been left, and there seemed little point in prolonging her misery by postponing her departure.

The flight over the North Pole was one of the most gruelling ever made by a solitary pilot. Once again Sheila had to fly without autopilot. After eight hours in the air on what was to be the crucial polar leg, wearing so many gloves that she could hardly move her fingers, she expected to be nearing the Pole. She could not think why her air speed was lower than expected until she noticed that the nose wheel had dropped partly out of its casing and the door was hanging down: at last she had discovered the reason for her earlier problems with speed and consumption. She would have to land, but her compasses were pointing in different directions when she made radio contact with Nord, a Danish weather station on an isolated island in a sea of ice. She neverthe-less managed to land safely, warmly welcomed by twenty-seven men who were delighted to meet the famous blonde pilot.

From Nord, she had no option but to fly in one leg to Barrow, which would take an estimated thirteen hours. To start with, the weather was clear, visibility as she approached the top of the world perfect, the frozen scenery beneath her impressively beautiful. Then she had to contend with ice, cloud and, worst of all, freezing rain. The physical effort of flying without autopilot was made worse by the intense cold inside the cabin, which at 9000 ft was – 9°C. Over the Pole, surrounded by mile upon mile of ice, she shouted triumphantly into her microphone: 'We're here – I'm on Top of the World!' Her only visual contact with life outside the cockpit was the NASA light which flashed to tell her that it was time for her mental acuity test: it was impossible to imagine that in Washington, the Goddard Space Center team monitoring her progress was sweltering in a heatwave. After seventeen hours alone in the Arctic air, Sheila relaxed in the warmth and hospitality of the Naval Arctic Research Laboratory at Barrow, where *Mythre*'s hydraulic trouble was traced and cured.

With her autopilot and wheel gear at last in working order, Sheila felt new confidence in *Mythre* as she flew down towards the Equator. From Australia, she flew as fast as possible back to England, fast enough to beat the record held by Jean Batten; but the pleasure of success was ruined by the return to her desecrated flat. Although exhausted, she was unable to resume a normal sleep pattern, and had moreover to make the difficult choice between *Myth Too* and *Mythre*. As she could not afford to keep both, she reluctantly sold the Comanche to a flying club.

In spite of her achievements in the air, Sheila had still not passed her driving test. A friend who had just become a driving instructor invited her to be his first pupil: her progress was followed in detail by the national press, and when she passed her test she was presented at the Biggin Hill Air Fair with a bright green Ford Capri just before she took off across the Atlantic to meet the men from Goddard, who had previously been merely names and voices. 'Anything that happens to me in the sky is fair enough and my life is worth losing for it if necessary, but it's the ground I fear,' she told an aviation correspondent prophetically before she left.

The weather on the flight to Washington was the roughest she had ever encountered, and after it *Mythre* had to be left at the Piper factory for repairs. Sheila had to wait for a broken rib to heal while she visited Goddard and accepted numerous invitations. When Hurricane Agnes hit North America, she was assured that her aircraft was safely out of the storm zone, only to discover that the airfield had been devastated and the neighbouring town virtually destroyed. *Mythre* was a shell, apparently unharmed from outside, but internally stripped and full of red slime, although Buck Rabbit had somehow survived. The bank, to which Sheila owed so much on the £50,000 aircraft, insisted that it be written off. The insurance company demanded a rebuild. Sheila, adamant that this should not be a patched-up job, fought a long and bitter battle, during which she was told that the only way she could recoup her financial losses was to sue those responsible for a misleading weather report. *Mythre* had been partially rebuilt when the bank called in the loan, and Sheila tried in vain to finish the repairs at her own expense. After over two years, she gave up the struggle, feeling that she had lost everything.

She still had her commercial flying licence, her Equity card and her fame as a writer, broadcaster and lecturer to fall back on, until an accident in the car she had been given before setting out on *Mythre's* last flight removed even her ability to work, leaving her with head injuries and almost blind. For more than two years she had one operation after another, gradually regaining her looks and her sight. In her mid-fifties, she had to start her life again from a basement flat in London, with no hope of realising her dream of flying and filming at the South Pole. Meanwhile, work based on the findings of her Arctic flight carried on: she had at least the comfort of knowing that she had contributed to search and rescue developments, to the understanding of the effects of stress and interrupted sleep patterns on pilots, and to the world battle against pollution of the atmosphere.

Although Sheila Scott flew with the advantage of modern technological developments, this did not make her flights any easier than those of the early pioneers who did not have to cope with their complexities. Her achievements as a modern pioneer pilot of great endurance and navigational skill were as great as theirs, often against the greater odds of the bad luck which dogged her on the ground and even on occasions in the air, where she felt after her long search for a meaning to life that she was at last 'on the very edge of understanding'.

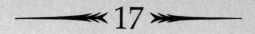

Conclusion:
Equality at Last?

When the men and women of the 1920s and 1930s made long-distance flights between and across continents, they were pioneering new routes and exploring the possibilities of aviation. Each record was a step towards modern air travel, whether it was a First (usually achieved by a man) between two places, a First between the same places by a woman, or a Fastest over a route already flown. The entertainment side of aviation contributed by bringing it to the public, the potential passengers, while competitions, with the emphasis on flying better and faster, spurred the manufacturers to work towards technical improvements.

The pioneering days have gone and can never return. Flying for fun, however, has continued alongside commercial and military aviation, the only restriction to membership of a club or ownership of a light aircraft now being the high cost of private flying. Fewer women than men belong to flying clubs, but there is nothing to stop those who can afford to and wish to fly from doing so. Women now participate equally with men in competitions, and fly with them or alone in balloons, gliders, light aircraft, hang gliders and microlights. They may also, if they can raise the money, emulate the pioneers by repeating and bettering their long-distance tests of endurance. From time to time, women have repeated the flights of the pioneers, following Amelia Earhart or Amy Johnson: they have however proved only that flying a long way in a light aircraft is still as strenuous and fraught with danger as ever, and have made no significant recent contribution to aviation through their self-inflicted endurance tests.

When Judith Chisholm flew alone round the world in 1980, halving Sheila Scott's record and breaking twenty-nine other records, she did it for fun; but she hoped nevertheless both that she would be able to recoup the costs of the venture and that she would cease to be discriminated against because she was a woman. She had learnt to fly in the 1960s as one of two female Air Traffic Controllers at Heathrow. Neither the RAF, nor the British Airways flying school at Hamble, would accept women for pilot training, but the four-year air traffic control training on a salary of £1100 a year included sufficient flying tuition for a private pilot's licence.

At her own expense Judith eventually gathered enough experience, while she worked for seven years in the control tower at Heathrow, to gain her instructor's and commercial licences, as well as an instrument rating, and flew as far as Africa in her 'little old Jodelle', a wooden fabric-covered aircraft with a tail wheel and upturned wings. In 1978 she started a two-year campaign for sponsorship for her solo flight round the world via Australia in a single-engined Cessna Centurion, but raising money for feats of endurance was even more difficult than in the early days of flying. She had eventually to borrow from a bank to put up over half the cost herself, although a national newspaper contributed £2000 and she was given £10,000 by a television company which featured her exploit in a short film: not a high price for what was expected to be an account of a glorious failure.

Judith, who set out to commemorate Amy Johnson's famous flight to Australia fifty years later, thought she would probably survive, but was not confident that she would break any records. Nevertheless she flew triumphantly round the world in 15 days, although she admitted to being scared over the ocean and hallucinating from lack of sleep. Unlike Amy Johnson, she enjoyed being a celebrity, but failed both to recover her costs and to find a sponsor for another long-distance record attempt: five years later, she was still trying to raise £100,000 to finance an assault on the men's 6½-day solo record from England to Australia.

Although Judith unashamedly used her femininity to seek sponsorship and publicity, she complained at the same time frequently and bitterly about discrimination against women in aviation. When she was successful in her first application as a commercial pilot, ferrying goods, personnel and VIP clients round Europe in executive jets, she found herself the only woman pilot based at Heathrow, and was incensed at the male attitude openly expressed by a senior pilot for another company when he turned her down for a freelance job with the explanation that the passengers 'would not like it if a little dolly bird started to fly the aeroplane'. She is outspoken about the conspiracy which she claims exists between the media and the airlines to discourage women pilots, and fears that organisations such as the British Women Pilots' Association are in danger of perpetuating the very discrimination which they seek to dispel.

Officially, applications to airlines by men and women pilots are now treated equally, both in Britain and elsewhere. In 1985, companies questioned directly about their policy on employing women pilots all claimed that they showed no discrimination against women, employed pilots on ability provided they satisfied basic requirements, and said that the small number of women pilots reflected the proportion of applicants. British Caledonian, for instance, employed two women out of a total of 340 pilots, and had 1000 applicants on their files of whom only six were women: their success would depend upon how they compared, in terms of both social and professional competence, with other interviewees.

Over half the British companies approached employed no women, although in theory they were willing to do so. British Airways had none, but at Air UK twelve of the pilot strength of just under two hundred were women, both the highest total and the biggest proportion. Air New Zealand had taken on three women pilots, in spite of the apocryphal story that a few years ago a chief pilot was prepared to hire women pilots provided they had passed the medical, had the right flying qualifications, and had played for the senior rugby team of a well-known boy's school. When the first women pilots were employed, the company felt it necessary to announce the fact: since then, however, it had been decided not to draw any particular attention to them as 'They are not women; they are pilots'. Germany, where the *Gleichhandlungsbesetz* – equal opportunities law – was passed only in 1981, still had no women flying for Lufthansa, the national airline.

One reason why so few women have applied in Britain for jobs as commercial pilots was simply that there was no way they should be trained except at their own expense. The British Airways flying school at Hamble took on one woman, in 1975, just before selling the school, and since then there has been no civil pilot training scheme in the country. In 1985 it cost over £30,000 to obtain a commercial pilot's licence, a sum which women found as difficult to raise as did the men. In 1986, British Airways announced a new pilot-training scheme to fill 100 positions in 1988. There were immediately thousands of inquiries, but only a very small proportion were from women. Britain's Civil Aviation Authority has no idea how many women have commercial licences, keeping its records encouragingly free of discrimination. The British Women Pilots' Association list of 300 members gives equally little indication of the number of professional female pilots, as its members include some who do not earn their living by flying, and some who do have never joined.

A large proportion of male commercial pilots in Britain still comes from the Royal Air Force, which specifically excludes women from pilot training on financial grounds: the RAF is 'not able to invest the large sums of money necessary to train women for fast jets'. Sixty years after they were set up, the University Air Squadrons opened their ranks to potential women pilots in July 1985: previously women could belong to the squadrons but could not avail themselves of the sponsored flying tuition available to male students. One woman medical officer has been trained by the RAF as a pilot, although she can wear only a Flight Medical Officer's badge and not RAF wings. The argument against training women as RAF pilots is that fewer men are likely to waste the cost of their training by leaving the service. There are still men in the RAF, however, who put forward, privately, the arguments that women's delicate internal organs will be damaged by high speed flying – apparently this applies only to female pilots, and not to passengers or stewardesses – and that G-suits do not fit them: they have, after all, been designed by men for men.

In France, it is still official Air Force policy to exclude women from flying jet aircraft on medical grounds. Acceleration is apparently not good for French women, in spite of the example of Jacqueline Auriol. In 1982, however, when two women were trained as army helicopter pilots, it was decreed that women should have equal access to training and employment in all except combat units; and in 1984, the first four women were accepted for French Air Force training as transport pilots. The condition that pilots must be of the male sex was dropped by Air France and the civil air training school, *l'Ecole Nationale de l'Aviation*

Civile, in 1973, but in the following ten years only 11 women availed themselves of the facility to become commercial pilots.

The United States seems the best country for a woman who wants to become a pilot. Out of a total of nearly 600,000 United States Air Force personnel 67,000, over one tenth, are not of the male sex, although this figure includes those in non-flying jobs, and women are as eligible as men for all pilot training except that specifically designed for combat. It is still a Federal Statute that women shall not engage directly in combat, although they can fly any size and speed of aircraft, including helicopters and jet trainers. By 1979, there were 110 women flying as pilots with commercial airlines in America, admittedly a minute proportion of the total of 45,000 commercial pilots, but one which has grown steadily since then. American women have also participated, on equal terms although in admittedly small numbers, in the exploration of space.

It was, however, a Russian, Valentina Tereshkova, who gained the title of first woman in space. She has since then become an ardent and long-winded spokeswoman for the Soviet Women's Committee and her country's drive for peace and international understanding between women and cosmonauts. She was twenty-six when, because of her parachuting experience with a club run by the factory at which she worked, she became a token woman in space during 48 orbital flights of the earth in spaceship Vostok VI in June 1963. Her presence added little more than prestige to the venture, although she became a Hero of the Soviet Union, married a cosmonaut, and toured both her own and other countries as a fervent and respected representative of Soviet womanhood.

It was not until five years after Valentina Tereshkova's space excursion that a woman, Yvonne Sintes of Dan-Air, was first appointed captain with a commercial airline, and it was 15 years before a western country sent a woman into space: in 1978, Dr Sally Ride was one of six women in an American team of 15 pilots and 20 mission specialists, and the only woman astronaut in the group to venture beyond the earth's atmosphere. Unlike the Soviet Union's space heroine, who left school with no qualifications and worked first in a tyre and then in a textile factory, America's spacewomen were highly qualified. In 1984 *Challenger* was sent into orbit with both Dr Sally Ride and Dr Kathryn

Sullivan, who was to be the first woman to walk in space during the mission, on board. A year later, yet another American woman astronaut, Dr Rhea Seddon, was one of a team of seven on space shuttle *Discovery* when it was sent into space to salvage the disabled communications satellite *Syncom*. Although the presence of women on American space missions received considerable publicity, they were appointed not as token women, nor with any self-consciousness about their sex, but because they happened to be the people best qualified for the job.

There were two American women in the crew of seven who were killed in the major space disaster in 1986. Judith Reznick, as a crew member, received little publicity either before or after the *Challenger* was launched into a cloudless blue February sky on its fated mission: the crew was usually listed by surname only, and equal opportunity in space for those with equal qualifications was by then unquestioned. It was Christa McAuliffe, the first 'ordinary member of the public' in space, who was the centre of attention. Mrs McAuliffe was not a token woman: she was a token person, chosen from thousands of applicants of both sexes as President Reagan's ideal representative of twentieth-century America. *Challenger*'s disintegration was watched by millions on television screens round the world. The stunned reaction of Christa McAuliffe's family was equally public. That the President's choice of person was a woman, a mother and a teacher, and was given an apple by her pupils just before take-off, and that her death was watched by her own children and those she taught, added poignancy to the disaster. No one suggested that her presence was in any way a contributory factor.

The space launch was an international television spectacle, and so this fatal space accident was spectacularly brought into private homes in a way which would have been unthinkable when air travel had reached the same stage as has space exploration in the 1980s. The exploration of space will, however, continue, as did aviation, in spite of accidents and deaths. It is ironic that women are already accepted as normal members of space crews, while they are still in such a tiny minority as airline pilots. Women are, however, as much to blame as men that there are still so few of them employed in aviation. They have been just as conditioned to accept a secondary role in most professions, or a domestic role rather than a profession, as have men to expect them to stay in their place, preferably on the ground or as stewardesses if they insist on taking to the air. Women pilots are now at least not expected to hand over to a subordinate man in the cockpit for announcements to passengers, in case the sound of a woman's voice should cause alarm. How many women might have taken to the air had it been easier for them to do so it is impossible to say: possibly not very many. It is thanks to those who in the past threequarters of a century have refused to stay on the ground that women are now, gradually, gaining equality in the air.

One whose equality was accepted without comment was thirty-four-year-old Jeana Yeager, partner and co-pilot in the non-stop, round-the-world flight of the *Voyager*, completed without refueling just before Christmas 1986. The 26,000-mile circumnavigation, hailed by President Reagan as the best Christmas present America could have wished for, did much to restore national morale in a year in which the public disaster of the *Challenger* had been followed by other failures in the space programme as well as by international political criticism of the country's foreign policy. It also made instant heroes of Dick Rutan, a forty-nine-year-old Vietnam veteran, and Jeana Yeager, an engineering and commercial draftsman with ten years' flying experience behind her and several previous world records, some of which Dick Rutan had previously held.

It was a personal rather than a national success, a tribute to the determination of three individuals: Jeana Yeager, Dick Rutan, and his brother Burt, designer of the aircraft. They had stuck doggedly to their conviction that the flight was possible, had raised two million dollars to pay for it, and had carried it out in spite of numerous setbacks. The ambition which Dick Rutan considered 'the last big plum in aviation history' had been shared equally from the beginning by Jeana. She has inevitably been compared to Amelia Earhart, while Dick Rutan is considered the modern equivalent of Charles Lindbergh. As they were living together during the five years they were working on the project, it was very much a partnership. Nine days of joint confinement in the *Voyager*'s cabin, which was just seven feet long and four feet high, must have tested their relationship as well as their stamina to the utmost. Sleep was almost impossible, as the ultra-light aircraft reacted strongly to any turbulance, however slight, and was on several occasions violently buffetted by storms. The constant noise of the engine threatened permanent loss of hearing, and increased the psychological and physical strain. Prepacked meals were heated on the radiator of the engine. As there was neither space nor money for a sophisticated waste-disposal system, solid waste had to be collected in bags and placed in a special wing compartment, and urine was vented through the fuselage through a funnel and tube. Although the two pilots managed occasional changes of underclothing, when they landed one of their first wishes was, not surprisingly, for a shower.

There were times when they wondered if they would manage to complete their ordeal, as when Jeana rubbed Dick's neck and shoulders and told him, 'We can get through. We can get through.' Her courage, he told reporters afterwards, was tremendous. Added to the discomfort was the knowledge that *Voyager* could easily have become a flying bomb had anything gone remotely wrong. With a wing span greater than that of a Boeing 727, it weighed only 2,000 lbs. and carried nearly four times its weight in fuel, stored in seventeen tanks in the wings and fuselage. Existing technology had been pushed to a logical extreme, which in turn imposed new demands on the stamina of the pilots.

Jeana Yeager, who is naturally shy and retiring, preferred whenever possible to leave radio contact during the flight and interviews after it to Dick. 'She just likes to get on with the job, and she is very much an equal partner in this,' was the comment from Mission Control at the Edwards Air Force Base. Although the project was aggressively marketed and had enthusiastic media coverage, there was no emphasis on the fact that one of the pilots was a woman, and Jeana did not receive any particular individual attention. One of the few facts to emerge about her was that she was not related to Chuck Yeager, first man to break the sound barrier; and, he, no doubt irritated by the constant repetition of the question as to their possibly being related, went on record as saying that the flight itself was of no more significance in the development of aviation than putting a big enough fuel tank into a car to drive it non-stop from Los Angeles to New York. Jeana, however, felt that they were 'taking a step into the future' and ushering in the next generation of aircraft.

Whether or not the *Voyager* flight proves to be a breakthrough remains to be seen. That it was a feat which combined technology with skill and endurance cannot be disputed. As Jeana commented shortly after landing, 'If it was easy, it would have been done a long time ago.' Her own professional and leisure interests, including such activities as skydiving, sailing, horse training and riding, skiing, backpacking, and jogging, had all no doubt helped to create the mental, physical, and technical fitness needed for the nine-day endurance test. The personal and national implications of successfully surviving the ordeal were recognized by the awarding of Presidential Citizen's Medals to the pilots and to the designer. 'History is still being written by men and women with a sense of adventure and derring-do,' commented President Reagan. Jeana Yeager was considered no more or no less of a modern-day pioneer, no more or no less newsworthy, because she was a woman: it was her achievement, rather than her sex, which was significant. Her predecessors would have approved.

When women pilots cease to become news, other than for outstanding achievements regardless of sex, the battle fought by pioneers such as Lady Heath, Amy Johnson, Amelia Earhart, Jacqueline Cochran and Jacqueline Auriol, and by many others whose names have been all but forgotten, will have been won. Perhaps one day there will no longer be newspaper headlines like those which have greeted almost every woman to earn her living as a pilot in the last few years. The American Brook Knapp who overcame her fear of flying to run her own successful aviation company was called the 'high-living fast lady'; Genair's five women pilots were 'women who are above it all'; when Lynn Ripplemeyer became the first woman captain of a jumbo, it was announced that 'Captain Lynn flies to place in history'; while 'I'm Betsy, let me fly you!' referred to the jumbo pilot Betsy Carroll. The assumption that a woman who flies is extraordinary is felt by feminists in aviation such as Judith Chisholm to be patronising, but there are still men who would agree with the writer of a letter printed in the magazine *Pilot* as recently as November 1984: 'Cannot the ladies leave entirely to the gentlemen one or two "worlds" – and I plead for aviation for one?'

The answer, from all women who fly or have flown, whether purely for pleasure, for sport or as a career, is an emphatic 'No'. From the earliest days of ballooning, women have shown their determination to participate in the pleasure and excitement of aviation. Time after time, they have proved their ability and endurance as pilots, sharing with their male colleagues a love of flying which has overcome convention and prejudice, and which will no doubt continue to do so. As Judith Chisholm has frequently said: 'All it takes is determination, an independent spirit and a thick skin' – qualities shown since the very beginning by women who have taken to the air.

Bibliography

* autobiographies

Andrews, Allen: *Back to the Drawing Board: the evolution of flying machines*. David & Charles, 1977.

*Auriol, Jacqueline: *I Live to Fly*. Flammarion (France), 1965; Michael Joseph, 1970.

Babington Smith, Constance: *Amy Johnson*. Collins, 1967

Backus, Jean: *Letters from Amelia*. Beacon Press (USA), 1982.

Bacon, Gertrude: *The Record of an Aeronaut*. John Long, 1907. * *Memories of Land and Sky*. Methuen & Co, 1938.

*Batten, Jean: *Alone in the Sky*. Airlife, 1979.

Bedford, John, Duke of: *The Flying Duchess*. Macdonald, 1968.

*Bird, Nancy: *Born to Fly*. Angus & Robertson, 1962.

Boase, Wendy: *The Sky's the Limit*. Macmillan, 1979.

Bostock, Peter: *The Great Atlantic Air Race*. Dent, 1970.

Brooks-Pazmanay, Kathleen: *United States Women in Aviation 1919–1929*. Smithsonian Studies in Air & Space (USA), 1983.

*Bruce, the Hon. Mrs Victor: *The Bluebird's Flight*. Chapman & Hall, 1931.

Burge, Squadron Leader C.G. (ed.): *Complete Book of Aviation*. Sir Isaac Pitman & Sons Ltd, 1935.

Burke, John: *Winged Legend*. Arthur Barker, 1970.

*Cochran, Jacqueline: *The Stars at Noon*. Little, Brown, 1954.

Cheesman, E.L.: *Brief Glory*. Harborough Publishing Co. Ltd, 1946.

Curtis, Lettice: *The Forgotten Pilots*. Foulis, 1971.

Davis, Burke: *Amelia Earhart*. G.P. Putnam & Sons (USA), 1972.

Dinesen, Isak: *Letters from Africa 1914–31*. Weidenfeld & Nicolson, 1982.

*Du Cros, Rosemary: *ATA Girl*. Muller, 1983.

Dwiggins, Don: *Flying Daredevils of the Roaring Twenties*. Arthur Barker, 1969.

*Earhart, Amelia: *20 hrs. 40 Mins*. G.P. Putnam & Sons (USA), 1929.
The Fun of It. Brewer, Warren & Putnam (USA), 1932.

Last Flight. Harcourt, Brace (USA), 1928.

Fox, James: *White Mischief*. Jonathan Cape, 1982.

Gibbs-Smith, Charles: *Ballooning*. Penguin, 1948.

Gore, J. (ed.): *Mary, Duchess of Bedford*. John Murray for Duke of Bedford, 1938.

*Gower, Pauline: *Women with Wings*. John Long, 1938.

Gunston, Bill (ed.): *Aviation: the Story of Flight*. Cathay Books, 1978.

The Plane Makers. Basinghall Books Ltd, 1980.

*Heath, Lady, and Murray, Stella Wolfe: *Women and Flying*. John Long, 1929.

Hudson, Kenneth and Pettifer, Julian: *Diamonds in the Sky*. Bodley Head, 1979.

Jablonski, Edward: *Atlantic Fever*. Macmillan Company, Inc. (USA), 1972.

Ladybirds: Women in Aviation. Hawthorn Books (USA), 1968.

Johnson, Amy: *Skyroads of the World*. W. & R. Chambers, 1939.

Myself when young, by famous women of today; ed. Margot, Countess of Oxford. Muller, 1938.

Lane, Peter: *Flight*. Batsford, 1974.

Lauwick, Hervé: *Heroines of the Sky*. Muller, 1960.

*Lindbergh, Anne Morrow: *Gift from the Sea*. Chatto & Windus, 1955.

North to the Orient. Harcourt, Brace & World, 1963.

Bring me a Unicorn: diaries and letters 1922–8. Chatto & Windus, 1972.

Hour of Gold, Hour of Lead: diaries and letters 1929–32. Chatto & Windus, 1973.

Locked Rooms and Open Doors: diaries and letters 1933–5. Chatto & Windus, 1974.

The Flower and the Nettle: diaries and letters 1936–5. Chatto & Windus, 1976.

*Markham, Beryl: *West with the Night*. Virago, 1984.

McKee, Alexander: *Into the Blue*. Granada Publishing, 1981.

Mondey, David (ed.): *Aviation: the complete book of aircraft and flight*. Octopus Books Ltd, 1980.

Moolman, Valerie: *Women Aloft*. Time-Life Books, 1981.

Mosley, Leonard: *Lindbergh*. Hodder & Stoughton, 1976.

Oakes, Claudia M. *United States Women in Aviation through World War I*. Smithsonian Studies in Air and Space (USA), 1978.

Palmer, Henry R.: *This was Air Travel.* Bonanza Books (New York), 1942.

Putnam, George Palmer: *Soaring Wings: a biography of Amelia Earhart.* Harcourt, Brace, 1939.

*Nicolson, Harold: *Diaries and Letters 1930–9.* Collins, 1966.

O'Donnell, James P.: *Berlin Bunker.* Arrow Books, 1980.

O'Neil, Paul: *Barnstormers and Speed Kings.* Time-Life Books, 1981.

*Reitsch, Hanna: *Fliegen mein Leben.* Deutschen Verlagsanstalt (Stuttgart), 1951 (translated as *The Sky My Kingdom*, The Bodley Head, 1955).

Ich flog fur Kwame Nkrumah. J.F. Lehmann (München), 1968.

Das Unzerstorbare in meinem Leben. Wilhelm Heyne Verlag (München), 1975

Höhen und Tiefen, ibid., 1978.

*Scott, Sheila: *On Top of the World.* Hodder & Stoughton, 1973.

*Shepherd, Dolly (with Peter Hearn and in collaboration with Molly Sedgwick): *When the 'chute went up. . . .* Robert Hale, 1984.

Strippel, Dick: *Amelia Earhart: the Myth and the Reality.* Exposition Press, 1972.

*Thaden, Louise: *High, Wide and Frightened.* Air Facts Press (USA), 1973.

Thurman, Judith: *The Life of Karen Blixen.* Weidenfeld & Nicolson, 1982.

Underwood, John: *The Stinsons.* Heritage Press (USA), 1969.

*Volkersz, Veronica: *The Sky and I.* W.H. Allen, 1956.

Villard, Henry: *Contact! The Story of the Early Birds.* Arthur Barker, 1968.

Waller, George: *Kidnap.* Hamish Hamilton, 1961.

*Welch, Ann: *Happy to Fly.* John Murray, 1983.

Wheeler, Air Commodore Alan: *Building Aeroplanes for 'Those Magnificent Men'.* Foulis, 1965.

*Yeager, Chuck: *Yeager.* Bantam Books (USA), 1985.

Index